PREFACE

This is a book about some of the ways in which the techniques of modal logic may be used to study concepts of proof theory first studied in Gödel's famous paper on the incompleteness of arithmetic. I have supposed that readers will have some knowledge of logic and at least a rough idea of what Gödel accomplished in his incompleteness paper, but not that they will know any modal logic. (Unfortunately, the lack of modal logic is easier to remedy than the lack of proof theory.)

An overview of the contents of the book is contained in the Introduction; the last paragraph of the Introduction explicitly lists the topics covered in the book. Here, I should like to acknowledge certain debts and certain shortcomings of the work.

My greatest debt is to Saul Kripke, who first told me about the system called '*G*' in this book. Kripke long ago realized the importance of *G* for proof theory, and proved the semantical completeness theorem for *G* established in Chapters 7 and 8. [Kripke also knew of the proof-theoretical importance of *S4Grz*, having proved that $\vdash_{S4Grz} A$ iff for all ϕ $\vdash_{PA} {}^{\phi}A$ (for all *A*) provided that $\vdash_G A$ iff for all ϕ $\vdash_{PA} A^{\phi}$ (for all *A*); cf. Chapter 13.]

Kripke has observed that there is an obvious extension of the treatment given here of modal propositional calculus to modal quantification theory, under which "in" quantifications have a totally unproblematical interpretation. Regrettably, the application of modal quantification theory to proof theory is not discussed in this book.

(Under Kripke's treatment, the converse Barcan formula is, but the Barcan formula itself is not, a theorem of the relevant system of modal quantification theory, the reason being that although every numerical instance of a provable universally quantified sentence of arithmetic is provable, unprovable universally quantified sentences exist all of whose instances are provable. One major open question in this area is whether the set of theorems of the relevant system is recursively enumerable.)

Robert Solovay was extremely generous with information about his own and related work. Solovay has announced a number of important theorems concerning interpretations of □ other than the ones discussed in this book, and I greatly regret that I have not been able to include these in the book. These theorems, as well as his completeness proofs for the modal systems *G* and *G**, are given in the *Israel Journal of Mathematics,* Volume 25 (1976), pages 287–304.

I am grateful to Hans van Maaren, the central argument of whose "Volledigheid v.d. modale logica *L*" enabled me to simplify the completeness proof for *G* given in Chapter 8, and to Cambridge University Press's referee for the care with which he went over the manuscript and for his long list of its errors and infelicities.

I am also grateful to Thomas Antognini, David Auerbach, Henk Barendregt, Paul Benacerraf, Claudio Bernardi, Richard Cartwright, Andrew Christie, Burton Dreben, Harvey Friedman, Robert Goldblatt, Warren Goldfarb, Harold Hodes, Richard Jeffrey, David Lewis, Ruth Marcus, Franco Montagna, Robert Nozick, Rohit Parikh, Charles Parsons, Jerzy Perzanowski, Hilary Putnam, W. V. O. Quine, Giovanni Sambin, Thomas Scanlon, Craig Smoryński, and Linda Wetzel.

And to Rebecca, Peter, and Bunch.

George Boolos

THE UNPROVABILITY OF CONSISTENCY

FOR MY MOTHER
AND
IN MEMORY OF MY FATHER

THE UNPROVABILITY OF CONSISTENCY

An essay in modal logic

GEORGE BOOLOS

Associate Professor of Philosophy, Massachusetts Institute of Technology

CAMBRIDGE UNIVERSITY PRESS

Cambridge

London · New York · Melbourne

CAMBRIDGE UNIVERSITY PRESS
Cambridge, New York, Melbourne, Madrid, Cape Town, Singapore, São Paulo, Delhi

Cambridge University Press
The Edinburgh Building, Cambridge CB2 8RU, UK

Published in the United States of America by Cambridge University Press, New York

www.cambridge.org
Information on this title: www.cambridge.org/9780521218795

First published 1979
This digitally printed version 2008

A catalogue record for this publication is available from the British Library

Library of Congress Cataloguing in Publication data

Boolos, George.

The unprovability of consistency.

Bibliography: p.

Includes index.

1. Modality (Logic) 2. Proof theory. I. Title.
BC199.M6B66 160 77-85710

ISBN 978-0-521-21879-5 hardback
ISBN 978-0-521-09297-5 paperback

CONTENTS

Preface vii *Introduction* 1

1 *G* and other normal modal propositional logics 19
2 Peano Arithmetic 34
3 The box as Bew 46
4 Some applications of *G* 61
5 Semantics for *G* and other modal logics 72
6 Canonical models 87
7 The completeness and decidability of *G* 98
8 Trees for *G* 108
9 Calculating the truth-values of fixed points 123
10 Rosser's theorem 133
11 The fixed-point theorem 141
12 Solovay's completeness theorems 151
13 An *S*4-preserving proof-theoretical treatment of modality 159
14 The Craig Interpolation Lemma for *G* 168
Notes 176
Bibliography 180
Index 183

INTRODUCTION

The purpose of this book is to show that modern modal logic is of interest to anyone interested in the concepts studied in Gödel's epochal paper "On formally undecidable propositions of *Principia Mathematica* and related systems I".[1] If modern modal logic was conceived in sin,[2] then it has been redeemed through Gödliness.

Modal logic

Modal logic is ordinarily conceived of as the study of the logical features of *necessity, possibility,* and related concepts, such as *implication* (necessity of the conditional). The notation of modal logic is that of ordinary logic, supplemented with the box □, which is usually read 'it is necessary that . . . ', or 'necessarily . . . '. The box may either be taken as a primitive sign or it may be defined from other signs not found in nonmodal logic, such as the diamond ◇, which is the possibility sign, or the fishhook ⥽, the implication sign. The usual interdefinitions are well known: '$\Diamond A$' may be regarded as an abbreviation of '$-\Box-A$'; '$\Box A$', of '$-\Diamond-A$'; '$(A ⥽ B)$', of '$\Box(A \rightarrow B)$'; and '$\Box A$', of '$((A \vee -A) ⥽ A)$'. Although there is an overwhelming diversity of systems of modal propositional logic that have been studied by logicians, most of these systems agree with one another with respect to the definition of (well-formed) sentence:

Except for notational variation, the same expressions count as sentences in almost all systems of modal propositional logic. It is in their accounts of *theorem,* or *asserted* sentence, that the systems differ from one another. We shall not speculate upon possible explanations of the diversity of systems.

We are going to use modal propositional logic to investigate the properties of the notions of provability and consistency in formal theories. We shall be particularly interested in the effects of taking the box of modal logic to mean 'it is provable that . . . ' rather than 'it is necessary that . . . '. (When the box is taken this way, the diamond comes to mean 'it is consistent that . . . '.) To study these effects we shall pay special attention to a particular system of modal propositional logic, called 'G', for Gödel. G has a number of familiar features. The same expressions that count as sentences in most modal propositional logics count as sentences of G. All tautologies and all sentences of the form $\Box(A \rightarrow A') \rightarrow (\Box A \rightarrow \Box A')$ are axioms of G. The rules of inference of G are the rules modus ponens and necessitation (from A, infer $\Box A$). (Substitution is a derived rule of G.) But because G has been devised to study the notion of provability rather than that of necessity, it has several unusual features. All sentences of the form,

$$\Box(\Box A \rightarrow A) \rightarrow \Box A,$$

are axioms of G, and indeed the only axioms of G other than those already mentioned. $\Box p \rightarrow p$ is thus *not* an axiom, nor is it even a theorem, of G. Moreover, no sentence of the form $\Diamond A$ is a theorem of G. On the other hand, $\Diamond p \rightarrow \Diamond(p \mathbin{\&} \Box -p)$ *is* a theorem of G.

C. I. Lewis, who founded modern modal logic, conceived of the subject as the study of the relation between propositions that is denoted by the word 'implies'. Lewis held that one proposition implies another just in case the other is *deducible* from the one. Exactly what Lewis meant by "deducible" may well be a question that cannot be answered with final certainty. Some passages in Lewis and Langford's *Symbolic Logic,* which was published almost twenty years after Lewis had published a paper on implication,[3] indicate that he sometimes had deducibility in deductive systems in mind when he spoke of implication:

> 17.32 $p \prec r \cdot q \prec s \cdot p \circ q : \prec \cdot r \circ s$ . . . For example, if a postulate p implies a theorem r, and a postulate q implies a theorem s, and the two postulates are consistent, then the theorems will be consistent. A system deduced from consistent postulates will be consistent throughout.[4]
>
> [With reference to *Principia Mathematica*]: When mathematical ideas have been defined – defined in terms of logical ideas – the postulates for arithmetic, such as Peano's postulates for arithmetic . . . can all be *deduced* [Lewis's italics].[5]
>
> In the light of all these facts, it appears that the relation of strict implication expresses precisely that relation which holds when valid deduction is possible, and fails to hold when deduction is not possible.[6]

But many other passages indicate that he meant the modal operators to be taken as referring to logical (metaphysical, mathematical) possibility – whatever that might be. For example:

> It should also be noted that the words "possible," "impossible," and "necessary" are highly ambiguous in ordinary discourse. The meaning here assigned to $\Diamond p$ is a *wide* meaning of "possibility" – namely, logical conceivability or the absence of self-contradiction . . . Since $p \prec q$ is defined in terms of logical impossibility as $\sim \Diamond(p \cdot \sim q)$, it is a narrow, or strict, meaning of "implies."[7]

However unclear Lewis may have been about the nature of the subject matter of the systems of modal logic that he created, he was certainly correct in thinking that deducibility is a concept that can and should be treated by the development of systems of modal logic similar to his own.[8] Deducibility and provability are strange notions, and different though their properties may be from those of implication and necessity, the symbolism of modal logic turns out to be an exceedingly useful notation for representing the forms of sentences of formal theories that have to do with the notions of deducibility, provability, and consistency, and the techniques devised to study systems of modal logic disclose facts about these notions that are of great interest.

The development of modal logic was greatly advanced when Kripke and others introduced certain versatile semantical notions into its study.[9] The box came to be seen as a kind of quantifier ranging over elements (often called "possible worlds" or just "worlds") of a nonempty set equipped with a binary relation. In Kripkean semantics sentences are evaluated as true or false *at* worlds: The truth-value at a world of a sentence letter (an atomic sentence) is specified outright; the truth-value at a world of a truth-functional compound is computed in the usual manner from the truth-values at that same world of its components; and a sentence $\Box A$ is true at a world if and only if A is true at all worlds to which the world bears the binary relation. A modal-logical model thus consists of a nonempty set (the domain), a binary relation on that set (the *accessibility* relation), and a function that assigns a truth-value to each pair consisting of a member of the domain and a sentence letter. Sentences are said to be *valid* in a model if they are true at all worlds of the model. Kripke demonstrated a number of soundness and completeness theorems about systems of modal logic of the form: The sentences that are theorems of (here follows the name of a well-known system) are precisely the sentences that are valid in all models in which the accessibility relation is (here follows an adjectival phrase for a well-known property of binary relations).

A system of modal propositional logic is called *normal* if the set of its theorems contains all tautologies and all sentences of the form $\Box(A \rightarrow A') \rightarrow (\Box A \rightarrow \Box A')$ and is closed under modus ponens, necessitation, and substitution. Like the better-known modal-logical systems T, $S4$, B, and $S5$, the system G is a normal system. And like those other systems, there is a Kripke-style soundness and completeness theorem that can be proved[10] for G: A relation R is called *well founded* if and only if there is no infinite sequence $w_0, w_1, w_2, w_3, \ldots$ such that $\cdots w_3Rw_2$ & w_2Rw_1 & w_1Rw_0. *The theorems of G are precisely the sentences valid in all models in which the converse of the accessibility relation is well founded and transitive.* (A relation is transitive if and only if its converse is.) Thus the theorems of G are the sentences valid in all models whose accessibility relation R is transitive and whose domain contains no $w_0, w_1, w_2, \ldots$ such that $w_0Rw_1Rw_2 \cdots$. The completeness theorem can be strengthened: 'Finite' may be inserted before 'models'. This strengthened completeness theorem for G, which is proved in Chapters 7 and 8, will be seen to be of great value for the study of provability and consistency, to which we now turn.

Arithmetic

In what follows, 'provable' and 'consistent' will mean 'provable in arithmetic' and 'consistent with arithmetic'. *Arithmetic,*[11] also known as *Peano Arithmetic* (*PA*), is (classical, first-order) formal arithmetic with induction and the usual axioms governing successor, addition, and multiplication. It is one of the two most-studied formal theories (Zermelo–Fraenkel set theory is the other), and is the formal theory with respect to which Gödel's incompleteness theorems are standardly proved. Facts analogous to those we shall establish about arithmetic can be established about many other formal theories, including Zermelo–Fraenkel set theory. But henceforth it is *arithmetic* that is under consideration.

In order to explain the connection between modal logic and arithmetic that is of interest to us, it is necessary to review certain notions and methods pertaining to PA, which were introduced by Gödel in "On formally undecidable propositions . . . ." (Gödel actually studied a system *P* different from, but closely related to, PA.)

We shall suppose that the reader knows what *Gödel numberings* are: mechanical (effective) one-one assignments of numbers to the expressions (finite sequences of symbols) of a language. Under such an assignment, the number assigned to an expression is called its *Gödel number*. We suppose the expressions of PA to have been assigned Gödel numbers in some familiar way.

The *numeral for* the number *n* is the expression of PA (it is a *term*) that is obtained when *n* occurrences of the successor sign are attached to the zero sign. In PA, the successor sign is $'$ and the zero sign is $\mathbf{0}$. Thus the numeral for three is $\mathbf{0}'''$ and that for zero is $\mathbf{0}$.

It will be convenient to have a notation for the numeral for the Gödel number of the expression *F* (of PA). We let '$\ulcorner F \urcorner$' denote this numeral.

As usual, we write '$\vdash_{\mathbf{PA}} F$' to mean that *F* is a theorem of PA.

Following Gödel's procedure, we can construct a formula Bew(x) (from *beweisbar*, "provable"), which satisfies the following three conditions[12]:

(I) If $\vdash_{\mathbf{PA}} S$, then $\vdash_{\mathbf{PA}} \mathrm{Bew}(\ulcorner S \urcorner)$;
(II) $\vdash_{\mathbf{PA}} \mathrm{Bew}(\ulcorner (S \to S') \urcorner) \to (\mathrm{Bew}(\ulcorner S \urcorner) \to \mathrm{Bew}(\ulcorner S' \urcorner))$;
(III) $\vdash_{\mathbf{PA}} \mathrm{Bew}(\ulcorner S \urcorner) \to \mathrm{Bew}(\ulcorner \mathrm{Bew}(\ulcorner S \urcorner) \urcorner)$
(for all sentences S, S' of PA).

The construction of Bew(x) is described in Chapter 2,

where sketches of the proofs of (I), (II), and (III) are also given.

We shall refer to Bew(⌜S⌝) as *the sentence that asserts that S is provable* (or that expresses the provability of S, etc.). According to the first condition, if a sentence S is provable, then so is the sentence that asserts that S is provable. According to the second condition, it is always provable in PA that if a conditional and its antecedent are provable, then so is its consequent. According to the third condition, it is always provable that a sentence S satisfies the *first* condition.

Condition (I) resembles the rule of necessitation in modal logic, and conditions (II) and (III) resemble the modal principles $\Box(A \rightarrow A') \rightarrow (\Box A \rightarrow \Box A')$ and $\Box A \rightarrow \Box\Box A$. This work is devoted to the exploitation of these resemblances.

The most obviously ingenious part of "On formally undecidable propositions . . . " was Gödel's construction of a sentence S *equivalent* (in P) *to the sentence that asserts that S is not provable.* The construction can be carried out in PA (and even in theories much weaker than PA). Moreover, the technique for constructing such an S is quite general, and can be used to show that for *any* predicate $P(x)$ of PA, there is a sentence S such that $S \leftrightarrow P(\ulcorner S \urcorner)$ is a theorem of PA.

A sentence S such that $S \leftrightarrow P(\ulcorner S \urcorner)$ is provable (in theory T) is called a *fixed point* (in T) of the predicate $P(x)$. Thus Gödel explicitly showed how to construct a fixed point (in P) of the predicate $-$Bew(x) [or, rather, of the analogue for P of $-$Bew(x)].

> We shall describe a general technique of fixed-point construction in Chapter 3, but a construction akin to Gödel's and applicable to a natural language (such as English) can be given here[13]: Let us recall that the result of enclosing a given (English) expression in a pair of quotation marks

is an expression that denotes (names, refers to), but is not identical to, the given expression. Thus, for example, the expression ''9'', which contains one, but not two, pairs of quotation marks, denotes the numeral '9', which is an expression that contains no quotation marks and denotes the number 9, which is not an expression at all.

We now show how to construct a sentence whose subject denotes *it* and whose predicate is, for example, 'is apt', which is a sentence that says of itself that it is apt.

For every expression *F* (of English), let *F* *normed* be the (English) expression that results when two occurrences of *F* are written down side by side and a pair of quotation marks is placed around the first occurrence. For example, 'is' normed = ''is' is', 'is apt' normed = ''is apt' is apt', and 'normed' normed = ''normed' normed'. (Thus 'normed' normed denotes itself.) And 'normed is apt' normed = ''normed is apt' normed is apt', which is a sentence whose subject, ''normed is apt' normed', denotes 'normed is apt' normed, which is that very sentence. 'Normed is apt' normed, therefore, is a sentence whose subject denotes it and whose predicate is 'is apt'.

A sentence of the language of a theory that is neither provable nor disprovable in the theory is called an *undecidable sentence* of the theory, or undecidable *in* the theory. (The notion of an undecidable sentence of a theory is not to be confused with that of an undecidable *theory*. There exist undecidable theories in which every sentence is decidable, and decidable theories in which there are undecidable sentences.)

A theory is called *incomplete* if there exist sentences that are undecidable in it.

Following Gödel, we can show that

(1) If PA is consistent, then no fixed point of $-\text{Bew}(x)$ is provable.

(2) If PA has a stronger property called "ω-consistency," then no fixed point of $-\text{Bew}(x)$ is disprovable.

Thus, since fixed points of $-\text{Bew}(x)$ exist,

(3) (Gödel's first incompleteness theorem for PA) If PA is ω-consistent, then PA is incomplete.

An improvement of Gödel's first incompleteness theorem was made in 1936 by Rosser, who showed that 'ω-' could be dropped from the statement of the theorem.[14] The nature and significance of Rosser's improvement are discussed in Chapter 10.

We can also show that

(4) Every conditional whose antecedent is the sentence of PA that expresses the consistency of PA and whose consequent is a fixed point of $-\text{Bew}(x)$ is provable. By (1), therefore,

(5) (Gödel's second incompleteness theorem for PA) If PA is consistent, then the sentence that expresses the consistency of PA is not provable.

The converse of (5) is evident, since every sentence of the language of an inconsistent theory is provable in that theory.

(1)–(5) are discussed at length in Chapters 2 and 3.

Which sentence is meant by 'the sentence of PA that expresses the consistency of PA'? There are at least four different, but coextensive, definitions of 'inconsistent theory' that are sometimes given: A theory may be said to be inconsistent, if every sentence of its language is provable in it, or if some sentence and its negation are provable in it, or if some particular sentence, whose negation is evidently a theorem, is provable, or if at least one contradiction is provable in it. An equally good definition, and the one we shall use, is to say that a theory is inconsistent if $\bot$ *is provable in it.* $\bot$ and $\top$ are the two 0-ary (i.e., 0-place) propositional connectives: $\bot$ is always evaluated as false; $\top$, as true. Negation is definable from $\bot$ and the conditional: $-p$ is equivalent to $(p \rightarrow \bot)$. We shall assume that we are discussing a version of PA in which $\bot$ is a primitive logical symbol, and we shall accordingly take the sentence of PA that expresses the consistency of PA to be

$$-\mathrm{Bew}(\ulcorner \bot \urcorner).$$

Thus (5) may be restated: If $\nvdash_{\mathrm{PA}} \bot$, then $\nvdash_{\mathrm{PA}} -\mathrm{Bew}(\ulcorner \bot \urcorner)$.

Up to now we have made little mention of semantic notions in our discussion of PA. We take them up now. *The standard model for PA* has for its universe (domain) the set of all natural numbers; **0** denotes zero in the standard model, and $'$, $+$, and $\cdot$ denote the successor, addition, and multiplication functions on the natural numbers. In the standard model, then, every numeral denotes the number it is the numeral for. The formula $\mathrm{Bew}(x)$ can be seen to be true (in the standard model) of a number n if and only if n is the Gödel number of a provable formula of PA. Thus, if S is a sentence of PA, then $\mathrm{Bew}(\ulcorner S \urcorner)$ *is true* (again in the standard model) if and only if *S is a provable sentence,* for $\mathrm{Bew}(\ulcorner S \urcorner)$ is true if and only if $\mathrm{Bew}(x)$ is true of the number denoted by $\ulcorner S \urcorner$, which is the Gödel number of S. Every theorem of PA is true in the standard model and therefore PA is consistent. Thus $-\mathrm{Bew}(\ulcorner \bot \urcorner)$ is true and therefore so is every fixed point of $-\mathrm{Bew}(x)$, for, by (4), each of these is deducible in PA from $-\mathrm{Bew}(\ulcorner \bot \urcorner)$.

Henkin raised the question of the status of fixed points of $\mathrm{Bew}(x)$, that is, of sentences S such that $S \leftrightarrow \mathrm{Bew}(\ulcorner S \urcorner)$ is a theorem of PA.[15] Each such sentence S is equivalent to the assertion that S is *provable.* Are they all provable (and hence true), are they all unprovable (and hence false), or are some provable and some unprovable? The answer to Henkin's question was given by Löb, who showed that for every sentence S, if $\mathrm{Bew}(\ulcorner S \urcorner) \rightarrow S$ is provable, then S itself is provable.[16] All sentences equivalent to their own provability are therefore provable and true. Löb's theorem for PA is proved in Chapter 3.

> It was not until sometime after Löb proved Löb's theorem (1954) that it was realized – perhaps first by Kripke[17] – that Löb's theorem is a direct consequence of Gödel's second incompleteness theorem (for single-sentence extensions of PA). Here is the argument:
>
> Let PA^+ be the result of augmenting PA with the axiom $-S$. PA^+ is consistent if and only if S is not provable in PA, and the sentence expressing the consistency of PA^+ is equivalent even in PA and, hence, in PA^+, to $-\text{Bew}(\ulcorner S \urcorner)$. Thus $\text{Bew}(\ulcorner S \urcorner) \rightarrow S$ is provable in PA if and only if $-S \rightarrow -\text{Bew}(\ulcorner S \urcorner)$ is provable in PA, if and only if $-\text{Bew}(\ulcorner S \urcorner)$ is provable in PA^+, if and only if the sentence expressing the consistency of PA^+ is provable in PA^+, if and only if, by the second incompleteness theorem for PA^+, PA^+ is inconsistent, if and only if S is provable in PA.
>
> Conversely, the second incompleteness theorem for PA immediately follows from Löb's theorem[18]: If $\nvdash_{PA} \perp$, then by Löb's theorem, $\nvdash_{PA} \text{Bew}(\ulcorner \perp \urcorner) \rightarrow \perp$, and so $\nvdash_{PA} -\text{Bew}(\ulcorner \perp \urcorner)$.

Modal logic and arithmetic

We can now describe the connection between modal logic and PA that will occupy us. We must define two notions: that of a realization and that of a translation. A *realization* is simply a function that assigns to each sentence letter of modal logic a sentence of arithmetic. The *translation A^ϕ of* a sentence A of modal logic *under* the realization ϕ is defined inductively:

> If A is a sentence letter, then A^ϕ is just $\phi(A)$.
> If $A = -B$, then $A^\phi = -(B^\phi)$.
> If $A = (B \,\&\, C)$, then $A^\phi = (B^\phi \,\&\, C^\phi)$.

Similarly, for all the other propositional connectives (so $\bot^\phi = \bot$).

$$\text{If } A = \Box B, \text{ then } A^\phi = \text{Bew}(\ulcorner B^\phi \urcorner).$$

(We take '$\Diamond$' to abbreviate '$-\Box-$'.)

So, for example, if $\phi(p) = \mathbf{0' + 0' = 0''}$, then $((\Box p \vee \Box - p) \,\&\, p)^\phi = ((\text{Bew}(\ulcorner \mathbf{0' + 0' = 0''} \urcorner) \vee \text{Bew}(\ulcorner - \mathbf{0' + 0' = 0''} \urcorner)) \,\&\, \mathbf{0' + 0' = 0''})$. And, no matter what ϕ may be, $(-\Box\bot \rightarrow -\Box - \Box\bot)^\phi = (-\text{Bew}(\ulcorner \bot \urcorner) \rightarrow -\text{Bew}(\ulcorner -\text{Bew}(\ulcorner \bot \urcorner) \urcorner))$, which is the sentence of PA that expresses the second incompleteness theorem for PA. If A is a sentence of modal logic and $A^\phi = S$, then $(\Box(\Box A \rightarrow A) \rightarrow \Box A)^\phi = \text{Bew}(\ulcorner (\text{Bew}(\ulcorner S \urcorner) \rightarrow S) \urcorner) \rightarrow \text{Bew}(\ulcorner S \urcorner)$, which is the sentence of arithmetic that asserts that S is provable if $\text{Bew}(\ulcorner S \urcorner) \rightarrow S$ is provable. Thus every translation of each axiom of G of the form $\Box(\Box A \rightarrow A) \rightarrow \Box A$ asserts that some particular instance of Löb's theorem holds. Each such translation, as we shall see, is itself provable, and indeed, *every translation of each theorem of G is a theorem of arithmetic.*

No system of which all translations of all theorems are theorems of arithmetic can have $(\Box p \rightarrow p)$ as one of its theorems: For if $\vdash_{\text{PA}} S \leftrightarrow -\text{Bew}(\ulcorner S \urcorner)$, then by the consistency of PA, $\nvdash_{\text{PA}} \text{Bew}(\ulcorner S \urcorner) \rightarrow S$ [otherwise the fixed point S of $-\text{Bew}(x)$ would be a theorem of PA, contra (1)]; thus, if $\phi(p)$ is a fixed point of $-\text{Bew}(x)$, then the translation of $(\Box p \rightarrow p)$ under ϕ is not a theorem of PA. [A second reason: If $\phi(p) = \bot$ and $\vdash_{\text{PA}} (\Box p \rightarrow p)^\phi$, then $\vdash_{\text{PA}} \text{Bew}(\ulcorner \bot \urcorner) \rightarrow \bot$, and so $\vdash_{\text{PA}} -\text{Bew}(\ulcorner \bot \urcorner)$, which, by the second incompleteness theorem, implies that PA is inconsistent.]

The absence of '$\Box A \rightarrow \Box\Box A$' from the description of the axioms of G has a different explanation: All sentences of this form are theorems of G (cf. Theorem 9 of Chapter 1); there is no need to take them as axioms.

By studying the modal logic G we can find out facts about provability and consistency in arithmetic that are of considerable interest. We shall consider a class of sentences that can be described, roughly, as the sentences "made up out of" Bew(x) and truth-functions (including $\bot$ and $\top$). The sentence of arithmetic that expresses the second incompleteness theorem is itself a member of this class, as are the assertions that arithmetic is consistent, that it is consistent with arithmetic that arithmetic is consistent, that it is provable that if arithmetic is consistent, then it is consistent that arithmetic is consistent, etc. We call these sentences *deictic* sentences. By examining the sentences of G that, like $-\Box\bot$, contain no sentence letters, we can see how to decide whether or not an arbitrary deictic sentence is true and how to decide whether or not an arbitrary deictic sentence is provable.[19]

Gödel, as we have noted, showed how to construct a sentence S such that $\vdash_{PA} S \leftrightarrow -\text{Bew}(\ulcorner S \urcorner)$, a sentence equivalent to the assertion that it is unprovable. Any such sentence is called a *Gödel sentence.* We shall study a class of sentences, which we call Gödelian fixed points, that contains every Gödel sentence, every sentence equivalent to the assertion that it is provable (i.e., every "Henkin sentence"), every sentence equivalent to the assertion that it is disprovable, every sentence equivalent to the assertion that it is disprovable if provable, etc. Gödel showed that all Gödel sentences are equivalent to the consistency assertion, which is a deictic sentence; Löb showed that all Henkin sentences are provable and hence equivalent to $\top$, which is also a deictic sentence. The Gödelian fixed points fall into natural classes, which can be defined in a simple manner by means of the notation of modal propositional logic and the notion of a realization (the Gödel sentences form one class, the Henkin sentences another, the sentences S that are equivalent to the assertion that S is disprovable if S is provable yet another, and so forth). In Chapter 9, we prove a theorem due to Bernardi and Smoryński that for every such class, there is a deictic sentence to which every member of the class is equivalent.[20] Thus the facts established by Gödel and Löb are two instances of a noteworthy regularity in arithmetic, which can and will be proved to hold by proving a metatheorem about G.

Perhaps the most interesting result that we shall prove is a theorem due to Robert Solovay, according to which the theorems of G are *precisely* those modal sentences of which all translations are theorems of arithmetic.[21] We call this theorem *Solovay's Completeness Theorem for G;* "completeness" here refers to realizations and not to models.

By reading the box as 'it is provable that . . . ', we can regard sentences of modal logic as expressing *principles of provability:* The principle expressed by a modal sentence is the (infinite) conjunction of all its translations. We may call a principle *provable* if all its conjuncts are provable, *correct* if all its conjuncts are true. Solovay's completeness theorem for G settles the question which sentences of modal logic express provable principles of provability.

Which sentences express correct principles of provability? Another completeness theorem of Solovay's settles this question as well. Every theorem of arithmetic is true and therefore, for every sentence S, Bew(⌜S⌝) → S is true. [By Löb's theorem, Bew(⌜S⌝) → S is *provable* if and only if S is provable; but whether or not S is provable, Bew(⌜S⌝) → S is always *true*.] Thus every translation of every sentence of the form $\Box A \rightarrow A$ is true. Every translation of every theorem of G is also true, since every theorem of arithmetic is true. And the truths of arithmetic are certainly closed under modus ponens. Thus every translation of every theorem of the (non-normal) system of modal logic whose axioms are all theorems of G and all sentences of the form $\Box A \rightarrow A$ and whose sole rule of inference is modus ponens is a truth. We call this system G*. Solovay's Completeness Theorem for G^* asserts that the sentences of modal logic that express correct principles of provability are precisely the theorems of G^*. Both of Solovay's completeness theorems are demonstrated in Chapter 12.

Lewis, we saw, supposed that there is some concept of deducibility under which fall precisely those pairs $\langle q,p \rangle$ of propositions such that the proposition with "antecedent" *p* and "consequent" *q* is necessary. Whether or not Lewis was correct, the properties of deducibility in arithmetic (provability of the conditional) differ sharply from the properties that the concept customarily signified by the fishhook is commonly supposed to have. Since $\Box\Diamond\top$ contains no sentence letters, all translations of $\Box\Diamond\top$ express the same proposition, namely the proposition that the consistency of arithmetic is provable in arithmetic.[22] This proposition is false, and $\Box\Diamond\top$ thus expresses an incorrect principle of provability in arithmetic (or in any consistent formalization of "total mathematics," for that matter). On the other hand, $\Box\Diamond\top \rightarrow \Box\bot$ expresses a correct principle of provability. To entertain the suggestion that $\Box\Diamond\top$ expresses an incorrect principle about mathematical or logical necessity, or that $\Box\Diamond\top \rightarrow \Box\bot$ expresses a correct one, for this or any other reason, strikes us as unwise, at best.

There is at least one philosophical misunderstanding that it is important to attempt to obviate, which has to do with W. V. O. Quine's critique of the notions of necessity and possibility. In a number of publications Quine has argued that we have no reason to believe that there are any statements that have the properties that it is commonly supposed that necessary truths have. If $\Box$ is read 'it is (logically, metaphysically, mathematically) necessary that . . . ', then it would be irrational or dogmatic of us to suppose that there are *any* truths of the form $\Box p$. According to Quine, for all anyone has been told, the box is a "falsum" operator. We do not wish to argue that logical necessity is a viable, respectable, intelligible, legitimate, or useful notion – or that it is not – but it is the purpose of this work to show that the mathematical ideas that have been invented to study this notion are interesting and useful. It may be *curious* that mathematical notions devised to study systems that are designed to study a notion of no philosophical or logical value should themselves be of considerable logical, philosophical, and mathematical interest, but it is surely not *impossible*. On the other hand, far from undermining Quine's critique of modal notions, the present work, if anything, helps to support Quine's position by supplying an example of an interpretation of the box whose intelligibility is beyond question. (The sin in which modal logic was said to be conceived is the confusion of use and mention. There is, of course, no such confusion in interpreting $\Box$ as 'it is provable that . . . ' in the manner we have described: If the translation of the modal sentence A is the sentence S of arithmetic, then the translation of $\Box A$ is Bew($\ulcorner S \urcorner$), the sentence that results when the numeral $\ulcorner S \urcorner$ for the Gödel number of S is substituted for the free variable x in the standard provability predicate Bew(x) of arithmetic, which is a sentence of arithmetic that is *perfectly well-defined*.)

The sort of connection between syntax and modality with which this work is concerned is to be contrasted with the sort of connection, or absence of one, that was established by Richard Montague in his "Syntactical treatments of modality . . . ."[23] Montague shows that no predicate $N(x)$ of any consistent extension T of Q (Robinson's arithmetic, a weak subtheory of arithmetic) will satisfy the following conditions for all sentences S, S' of T:

(A) $\vdash_T N(\ulcorner S \urcorner) \rightarrow S$,
(B) $\vdash_T N(\ulcorner N(\ulcorner S \urcorner) \rightarrow S) \urcorner)$,
(C) $\vdash_T N(\ulcorner (S \rightarrow S') \urcorner) \rightarrow (N(\ulcorner S \urcorner) \rightarrow N(\ulcorner S' \urcorner))$, and
(D) $\vdash_T N(\ulcorner S \urcorner)$, if S is a logical axiom.

Montague concludes that "if necessity is to be treated syntactically, that is, as a predicate of sentences, . . . then virtually all of modal logic must be sacrificed." Montague's concern was to show that neither Bew(x) nor any other predicate of any sufficiently strong theory possesses all of the most elementary modal-logical features of the sentential operator 'necessarily'. Montague may well have established the impossibility of simultaneously treating modal notions *syntactically* and maintaining the customary laws of modal logic. But our interest is to consider what laws do arise when the box is treated syntactically, by taking it to mean Bew(⌜ ⌝), and to show the utility of a modal treatment of syntactic notions.

Montague's paper suggests a question that may have an interesting answer: What other arithmetical notions and modal systems besides Bew(x) and G are there for which a Solovay-style completeness theorem can be proved? One answer is given in Chapter 13.

We now briefly describe the contents of the rest of this book. We first review some familiar modal systems and contrast them with G; then we review those facts about Peano Arithmetic that we shall need later on. We then show that every translation of every theorem of G is a theorem of arithmetic and discuss some familiar theorems about arithmetic in the light of G. We then prove some recent results about deictic sentences and about Gödelian fixed points and provability. The standard semantics for systems of modal logic is presented, and we show that the theorems of G are valid in models in which the accessibility relation is transitive and has a well-founded converse. Canonical models are explained, and the completeness of certain well-known modal systems is demonstrated by means of canonical models before that of G is so demonstrated. A practical and pleasant method – a tree method – for working with G is introduced, and a totally different completeness proof for G is given via trees. The completeness theorem is used to prove the Bernardi–Smoryński theorem on Gödelian fixed points. We then digress to discuss connections between Rosser's theorem and G. We next present the raisons d'être of this book: the beautiful generalization by de Jongh and Sambin of the Bernardi–Smoryński theorem and Solovay's completeness theorems for G and G^*. We then give a proof-theoretical treatment of modality in which all the theorems of $S4$, including $(\Box p \rightarrow p)$, are upheld. And in the final chapter we prove the Craig Interpolation Lemma for G.

1

G and other normal modal propositional logics

We are going to investigate a system of modal propositional logic, which we call 'G', for Gödel. By studying G we can learn new and interesting facts about *provability, consistency,* and *self-reference,* the concepts studied by Gödel in his paper "On formally undecidable propositions of *Principia Mathematica* and related systems I."[1]

Like the systems T (sometimes called 'M'), $S4$, B, and $S5$, which are perhaps the four best-known systems of modal logic, G is a *normal* system of modal propositional logic. That is to say, the set of theorems of G contains all tautologies of the propositional calculus (including, of course, those that may contain the special symbols of modal logic), contains all distribution axioms, that is, sentences of the form $\Box(A \to A') \to (\Box A \to \Box A')$, and is closed under the rules modus ponens, necessitation [infer (the theoremhood of) $\Box A$ from (the theoremhood of) A], and substitution. Nor does G differ from those other systems in the syntax of its sentences: A sequence of "symbols" is a sentence of T, $S4$, B, or $S5$ if and only if it is a sentence of G.

But G differs greatly from T, $S4$, B, and $S5$ with respect to basic questions of theoremhood. In addition to tautologies and distribution axioms, its other axioms are the substitution instances of $\Box(\Box p \to p) \to \Box p$, that is, the sentences of the form $\Box(\Box A \to A) \to \Box A$, and it has no other rules of inference than the ones already mentioned. Now the sentence p is a tautological consequence of $\Box(\Box p \rightarrow p) \rightarrow \Box p$ and two extremely familiar principles of modal logic, namely $\Box p \rightarrow p$ and $\Box(\Box p \rightarrow p)$, both of which are theorems of all of T, $S4$, B, and $S5$. Since G is a normal system of modal propositional logic, $\Box p \rightarrow p$ cannot be one of its theorems, on pain of inconsistency: For if $\Box p \rightarrow p$ were a theorem, then by necessitation $\Box(\Box p \rightarrow p)$ would also be a theorem, and then, by two applications of modus ponens, p would also be a theorem, whence by substitution, $\bot$ would be a theorem, and G would be inconsistent. However, G is perfectly consistent and $\Box p \rightarrow p$, as we shall see in this chapter, is *not* one of its theorems.

In order to compare and contrast G with its better-known but less deserving relatives, we shall take a general look at systems of modal propositional logic. Much of the material in this chapter will be quite familiar, but it will be important to reverify certain elementary facts, in order to establish that they hold in the absence of $\Box p \rightarrow p$, which we shall be living without in most of this book. The material in this chapter will be of a purely "syntactic" or "proof-theoretical" character. We shall not take up the semantics of modal logic until Chapter 5.

We begin our general look at modal logic by defining the notion of a *modal sentence*. (For the sake of brevity, "modal" will sometimes be omitted before "sentence.")

We select a countably infinite sequence of distinct objects, of which the first five are called '$\bot$', '$\rightarrow$', '$\Box$', '(', and ')', and the others are called "sentence letters". Modal sentences will be certain finite sequences of these objects. 'p', 'q', 'p''', 'p_{17}', etc., will be used as variables ranging over sentence letters, under the usual convention that in the same context, distinct variables for sentence letters range over distinct sentence letters. If A and B are finite sequences of objects, then AB is the result of appending B to A. (We do not here need to distinguish between an object and the sequence consisting just of that object.)

Here, then, is the (inductive) definition of *modal sentence*.

(1) $\bot$ is a modal sentence.
(2) Each sentence letter is a modal sentence.
(3) If A and B are modal sentences, then so is $(A \to B)$.
(4) If A is a modal sentence, then so is $\Box A$.

$\Box A$ is the *necessitation of A*.

Sentences that do not contain sentence letters are *letterless*. For example, $\bot$, $\Box\bot$, and $(\Box\bot \to \bot)$ are letterless sentences.

Some explanation may be in order for our curious choice of the 0-ary propositional connective $\bot$ as a primitive symbol. First of all, $\bot$, which is always evaluated as false in the propositional calculus, and $\to$ together form a complete set of propositional connectives in the sense that every propositional connective can be defined from them, for $-A$, the negation of A, is equivalent to $(A \to \bot)$, and, as is well known, $-$ and $\to$ form a complete set of connectives. Second, if $\bot$ is taken as a primitive symbol, then a handy, completely general, and nonarbitrary way to say that a system is consistent is simply to say that $\bot$ is not one of its theorems. Third, taking $\bot$ as a primitive guarantees that there will be some letterless sentences; these will later turn out to be convenient "representations" of certain interesting propositions expressible in the language of arithmetic. For example, $\Box\bot$ will "represent" the (false) proposition that arithmetic is inconsistent, and $-\Box\bot \to -\Box-\Box\bot$, the second incompleteness theorem.

Here are some familiar definitions:

$$-A =_{df} (A \rightarrow \bot);$$
$$\top =_{df} -\bot;$$
$$(A \vee B) =_{df} (-A \rightarrow B);$$
$$(A \,\&\, B) =_{df} -(A \rightarrow -B);$$
$$(A \leftrightarrow B) =_{df} (A \rightarrow B) \,\&\, (B \rightarrow A);$$
$$\Diamond A =_{df} -\Box -A.$$

We shall later need the notion of a *subsentence* of a sentence. Its (inductive) definition is: Every sentence is a subsentence of itself; if $(B \rightarrow C)$ is a subsentence of A, then B and C are subsentences of A; if $\Box B$ is a subsentence of A, then B is a subsentence of A.

We shall take a *modal propositional calculus* to be a set of sentences, called the *axioms* of the calculus, together with a set of relations on sentences, called the *rules of inference* of the calculus. As usual, a proof of a sentence in a calculus in a finite sequence of sentences, each of which either is an axiom of the calculus or is immediately deducible from earlier sentences in the sequence by one of the rules of inference, and the last of which is the sentence. (B is said to be *immediately deducible from* $A_1, \ldots, A_n$ by the rule of inference R if $\langle A_1, \ldots, A_n, B\rangle \in R$.) And a sentence is a theorem of a calculus if there is a proof of the sentence in the calculus. ('$\vdash_L A$' means 'A is a theorem of L'.)

Modus ponens is the relation containing all triples $\langle A, (A \rightarrow B), B\rangle$.

Necessitation is the relation containing all pairs $\langle A, \Box A\rangle$.

Substitution is the relation containing all pairs $\langle A, B\rangle$ such that for some sentence letters $p_1, \ldots, p_n$ and some sentences $C_1, \ldots, C_n$, B is the result of respectively replacing every occurrence of $p_1, \ldots, p_n$ in A by an occurrence of $C_1, \ldots, C_n$. If B is immediately deducible from A by substitution, then B is called a *substitution instance* of A.

A *distribution axiom* is a sentence "of the form"

$(\Box(A \rightarrow A') \rightarrow (\Box A \rightarrow \Box A'))$, that is, a sentence which *is* $(\Box(A \rightarrow A') \rightarrow (\Box A \rightarrow \Box A'))$, for some sentences A and A'.

We now present seven modal propositional calculi. All tautologies and all distribution axioms are axioms of each system, and the rules of inference of each of them are just modus ponens and necessitation. *The calculus K* has no other axioms. The other axioms of *the calculus K4* are the sentences of the form $\Box A \rightarrow \Box\Box A$. The other axioms of *the calculus T* are the sentences of the form $\Box A \rightarrow A$. The other axioms of *the calculus S4* are the sentences of the forms $\Box A \rightarrow A$ and $\Box A \rightarrow \Box\Box A$. The other axioms of *the calculus B* are the sentences of the forms $\Box A \rightarrow A$ and $A \rightarrow \Box\Diamond A$. The other axioms of *the calculus S5* are the sentences of the forms $\Box A \rightarrow A$ and $\Diamond A \rightarrow \Box\Diamond A$.

The other axioms of *the calculus G* are the sentences of the form $\Box(\Box A \rightarrow A) \rightarrow \Box A$.

One system is said to be *extended by* another if every theorem of the one is a theorem of the other. If we abbreviate 'every theorem of L is a theorem of M' by '$L \rightleftharpoons M$', then we evidently have

$$\begin{array}{ccccc} & & G & & \\ & & \Uparrow & & \\ & & K & \rightleftharpoons & K4 \\ & & \Downarrow & & \Downarrow \\ S5 & \leftrightharpoons & T & \rightleftharpoons & S4 \\ & & \Downarrow & & \\ & & B & & \end{array}$$

By the end of the chapter we shall have shown that, in fact,

$$\begin{array}{ccccc} K & \rightleftharpoons & K4 & \rightleftharpoons & G \\ \Downarrow & & \Downarrow & & \\ T & \rightleftharpoons & S4 & & \\ \Downarrow & & \Downarrow & & \\ B & \rightleftharpoons & S5 & & \end{array}$$

But our first task will be to verify that the seven systems are all normal. Since the axioms of each include all tautologies and all distribution axioms and the rules of each are just modus ponens and necessitation, it suffices to show that every substitution instance of a theorem of one of the systems is also a theorem of that system.

Suppose then that A is a theorem of one of the systems, and that B is the result of respectively replacing all occurrences in A of the sentence letters $p_1, \ldots p_n$ by occurrences of the sentences $C_1, \ldots, C_n$. Let $A_1, \ldots, A$ be a proof of A in the system. For every sentence D, let D^o be the result of replacing every occurrence of one of the ps in D by an occurrence of the corresponding C. Then $A^o = B$. We want to see that $A_1{}^o, \ldots, A^o$ is a proof of B. To see this, observe that every substitution instance of an axiom of one of the systems is also an axiom of the system, and that if A_k is immediately deducible from A_i and A_j by modus ponens, or from A_i by necessitation, then, since $(D \rightarrow E)^o = (D^o \rightarrow E^o)$ and $(\Box D)^o = \Box D^o$, the same holds for $A_k{}^o$, $A_i{}^o$, and $A_j{}^o$.

Our seven systems are thus normal systems of modal propositional logic. (We have *not* followed Kripke's "Semantical analysis of modal logic I: Normal modal propositional calculi"[2] in requiring that all substitution instances of $\Box p \rightarrow p$ be theorems of a system if that system is to be called 'normal'. This departure is by now customary.)

Normal systems are closed under truth-functional consequence: If B is a truth-functional consequence of the theorems $A_1, \ldots, A_n$ of a normal system, then B *is also a theorem of the system, for* B can be inferred by n applications of modus ponens from $A_1, \ldots, A_n$ and the tautology $(A_1 \rightarrow (\cdots (A_n \rightarrow B) \cdots))$.

Moreover, if $A \rightarrow B$ is a theorem of a normal system L, then so is $\Box A \rightarrow \Box B$. For if $\vdash_L A \rightarrow B$, then by necessitation, $\vdash_L \Box(A \to B)$. But since $\Box(A \to B) \to (\Box A \to \Box B)$ is a distribution axiom, by modus ponens, $\vdash_L \Box A \to \Box B$. Consequently, if $A \leftrightarrow B$ is a theorem of a normal system, then so is $\Box A \leftrightarrow \Box B$. K (for 'Kripke') is the smallest normal system of modal propositional logic in the sense that it is extended by every other normal system. Two important facts about K are stated in the next two theorems.

Theorem 1
$\vdash_K \Box(A \mathbin{\&} B) \leftrightarrow (\Box A \mathbin{\&} \Box B)$ (for all sentences A, B).

Proof. Since $(A \mathbin{\&} B) \to A$ is a tautology, $\vdash_K (A \mathbin{\&} B) \to A$, and so $\vdash_K \Box(A \mathbin{\&} B) \to \Box A$. Similarly, $\vdash_K \Box(A \mathbin{\&} B) \to \Box B$, and so by the propositional calculus, $\vdash_K \Box(A \mathbin{\&} B) \to (\Box A \mathbin{\&} \Box B)$. For the converse, observe that $\Box(B \to (A \mathbin{\&} B)) \to (\Box B \to \Box(A \mathbin{\&} B))$ is a distribution axiom and hence a theorem of K. But then, since $\vdash_K A \to (B \to (A \mathbin{\&} B))$, $\vdash_K \Box A \to \Box(B \to (A \mathbin{\&} B))$, and so by the propositional calculus $\vdash_K \Box A \to (\Box B \to \Box(A \mathbin{\&} B))$. We thus have $\vdash_K (\Box A \mathbin{\&} \Box B) \to \Box(A \mathbin{\&} B)$ and therefore $\vdash_K \Box(A \mathbin{\&} B) \leftrightarrow (\Box A \mathbin{\&} \Box B)$. ⊣

Theorem 2
$\vdash_K \Box(A_1 \mathbin{\&} \cdots \mathbin{\&} A_n) \leftrightarrow (\Box A_1 \mathbin{\&} \cdots \mathbin{\&} \Box A_n)$ (for all sentences $A_1, \ldots, A_n$).

Proof. Use Theorem 1 $n - 1$ times. ⊣

Theorem 3
Suppose that L is a normal system. If $\vdash_L A_1 \mathbin{\&} \cdots \mathbin{\&} A_n \to B$, then $\vdash_L \Box A_1 \mathbin{\&} \cdots \mathbin{\&} \Box A_n \to \Box B$.

Proof. If $\vdash_L A_1 \mathbin{\&} \cdots \mathbin{\&} A_n \to B$, then since L is normal, $\vdash_L \Box(A_1 \mathbin{\&} \cdots \mathbin{\&} A_n) \to \Box B$. By Theorem 2, $\vdash_K \Box(A_1 \mathbin{\&} \cdots \mathbin{\&} A_n) \leftrightarrow (\Box A_1 \mathbin{\&} \cdots \mathbin{\&} \Box A_n)$. Since K is extended by L, $\vdash_L \Box A \mathbin{\&} \cdots \mathbin{\&} \Box A_n \to \Box B$. ⊣

Corollary
If L is a normal system, then $\vdash_L \Diamond A \,\&\, \Box B \rightarrow \Diamond(A \,\&\, B)$.

Proof. $\vdash_L B \,\&\, -(A \,\&\, B) \rightarrow -A$. By Theorem 3, $\vdash_L \Box B \,\&\, \Box-(A \,\&\, B) \rightarrow \Box-A$. Contraposing, we have $\vdash_L \Diamond A \,\&\, \Box B \rightarrow \Diamond(A \,\&\, B)$. ⊣

Theorem 4
Suppose that $F(A)$ $(F(B))$ is the result of replacing every occurrence of p in some sentence F by an occurrence of the sentence A (B). Then, if L is a normal system and $\vdash_L A \leftrightarrow B$, we have $\vdash_L F(A) \leftrightarrow F(B)$.

Proof. The proof is an induction on the complexity of sentences. Suppose that $\vdash_L A \leftrightarrow B$. If F is $\perp$, then $F(A)$ is $\perp$ and $F(B)$ is $\perp$. Since $\perp \leftrightarrow \perp$ is a tautology, we have $\vdash_L F(A) \leftrightarrow F(B)$. If F is p itself, then $F(A)$ is A and $F(B)$ is B. If F is some sentence letter q other than p, then $F(A)$ and $F(B)$ are both q. But $\vdash_L q \leftrightarrow q$. Suppose that F is $(G \rightarrow H)$. Then $F(A)$ is $(G(A) \rightarrow H(A)$ and $F(B)$ is $(G(B) \rightarrow H(B))$. We may assume that $\vdash_L G(A) \leftrightarrow G(B)$ and $\vdash_L H(A) \leftrightarrow H(B)$. But then by the propositional calculus, $\vdash_L F(A) \leftrightarrow F(B)$. Finally, suppose that F is $\Box G$. Then $F(A)$ is $\Box G(A)$ and $F(B)$ is $\Box G(B)$. We may assume that $\vdash_L G(A) \leftrightarrow G(B)$. But then, since L is normal, $\vdash_L \Box G(A) \leftrightarrow \Box G(B)$. ⊣

Theorems 3 and 4 will frequently be used without explicit mention in what follows. In particular, we shall freely interchange $\Box$ and $-\Diamond-$, $-\Box$ and $\Diamond-$, and $-\Diamond$ and $\Box-$, and truth-functionally equivalent sentences.

An important property of $K4$ is stated in Theorem 5.

Theorem 5
Same supposition as in Theorem 4. Then $\vdash_{K4} \Box(A \leftrightarrow B) \rightarrow \Box(F(A) \leftrightarrow F(B))$.

Proof. The proof is the formalization in $K4$ of the proof of Theorem 4. If F is $\bot$, then since $\vdash_{K4} \bot \leftrightarrow \bot$, by necessitation, $\vdash_{K4} \Box(\bot \leftrightarrow \bot)$, and so $\vdash_{K4} \Box(A \leftrightarrow B) \rightarrow \Box(\bot \leftrightarrow \bot)$. If F is p, then $\vdash_{K4} \Box(A \leftrightarrow B) \rightarrow \Box(A \leftrightarrow B)$. If F *is* q, $\neq p$, then $\vdash_{K4} \Box(A \leftrightarrow B) \rightarrow \Box(q \leftrightarrow q)$. Suppose that F is $(G \rightarrow H)$. By the hypothesis of the induction, we may assume that $\vdash_{K4} \Box(A \leftrightarrow B) \rightarrow \Box(G(A) \leftrightarrow G(B))$ and $\vdash_{K4} \Box(A \leftrightarrow B) \rightarrow \Box(H(A) \leftrightarrow H(B))$. But then, since $(F(A) \leftrightarrow F(B))$ is a truth-functional consequence of $(G(A) \leftrightarrow G(B))$ and $(H(A) \leftrightarrow H(B))$, by the propositional calculus and Theorem 3, $\vdash_{K4} \Box(A \leftrightarrow B) \rightarrow \Box(F(A) \leftrightarrow F(B))$. Finally, suppose that F is $\Box G$. We may assume that $\vdash_{K4} \Box(A \leftrightarrow B) \rightarrow \Box(G(A) \leftrightarrow G(B))$. Since each of $G(A)$ and $G(B)$ is a truth-functional consequence of $(G(A) \leftrightarrow G(B))$ and the other, by the propositional calculus and Theorem 3, we have $\vdash_{K4} \Box(A \leftrightarrow B) \rightarrow (F(A) \leftrightarrow F(B))$. As $K4$ is normal, $\vdash_{K4} \Box\Box(A \leftrightarrow B) \rightarrow \Box(F(A) \leftrightarrow F(B))$. And since $\vdash_{K4} \Box(A \leftrightarrow B) \rightarrow \Box\Box(A \leftrightarrow B)$, we have $\vdash_{K4} \Box(A \leftrightarrow B) \rightarrow \Box(F(A) \leftrightarrow F(B))$ in this case too. $\dashv$

We want to emphasize that Theorem 5 is a theorem about $K4$ and hence about all extensions of $K4$. No use of $\Box p \rightarrow p$ has been made in its proof.

Theorem 6

$\vdash_T A \rightarrow \Diamond A$; $\vdash_T \Box A \rightarrow \Diamond A$.

Proof. $\vdash_T \Box -A \rightarrow -A$. Contraposing, we obtain $\vdash_T A \rightarrow \Diamond A$. Since $\vdash_T \Box A \rightarrow A$, $\vdash_T \Box A \rightarrow \Diamond A$. $\dashv$

In the next two theorems we demonstrate two very familiar facts about the systems $S4$, B, and $S5$. We then proceed to demonstrate some basic facts about G. We shall make frequent use of the fact that if L is a normal system, then $\vdash_L \Diamond A \rightarrow \Diamond B$ if $\vdash_L A \rightarrow B$, for then $\vdash_L -B \rightarrow -A$, $\vdash_L \Box -B \rightarrow \Box -A$, and $\vdash_L -\Box -A \rightarrow -\Box -B$, that is, $\vdash_L \Diamond A \rightarrow \Diamond B$.

Theorem 7
$\vdash_{S4} \Box\Diamond A \leftrightarrow \Box\Diamond\Box\Diamond A$.

Proof. By Theorem 6, $\vdash_T \Box\Box\Diamond A \rightarrow \Diamond\Box\Diamond A$, and so by normality,

$\vdash_T \Box\Box\Box\Diamond A \rightarrow \Box\Diamond\Box\Diamond A$. But
$\vdash_{S4} \Box\Diamond A \rightarrow \Box\Box\Diamond A$ and
$\vdash_{S4} \Box\Box\Diamond A \rightarrow \Box\Box\Box\Diamond A$. Since T is extended by $S4$,
$\vdash_{S4} \Box\Diamond A \rightarrow \Box\Diamond\Box\Diamond A$. Conversely, by Theorem 6
$\vdash_T \Box\Diamond A \rightarrow \Diamond\Diamond A$, and so by normality,
$\vdash_T \Diamond\Box\Diamond A \rightarrow \Diamond\Diamond\Diamond A$, and by normality again,
$\vdash_T \Box\Diamond\Box\Diamond A \rightarrow \Box\Diamond\Diamond\Diamond A$. But since
$\vdash_{S4} \Box -A \rightarrow \Box\Box -A$,
$\vdash_{S4} \Diamond\Diamond A \rightarrow \Diamond A$, and so by normality,
$\vdash_{S4} \Diamond\Diamond\Diamond A \rightarrow \Diamond\Diamond A$, and then
$\vdash_{S4} \Diamond\Diamond\Diamond A \rightarrow \Diamond A$, whence by normality,
$\vdash_{S4} \Box\Diamond\Diamond\Diamond A \rightarrow \Box\Diamond A$. Since T is extended by $S4$,
$\vdash_{S4} \Box\Diamond\Box\Diamond A \rightarrow \Box\Diamond A$. ⊣

We shall see that $\vdash_G \Box\Diamond A \leftrightarrow \Box\Diamond\Box\Diamond A$ also, for a quite different reason.

We now want to prove that a normal system extends $S5$ if and only if it extends both $S4$ and B. It will suffice to show that $S5$ has the same theorems as the normal modal propositional calculus – we shall call it '$5S$' – whose additional axioms are all sentences $\Box A \rightarrow A$, $\Box A \rightarrow \Box\Box A$, and $A \rightarrow \Box\Diamond A$, and whose rules of inference are modus ponens and necessitation.

Theorem 8
$\vdash_{S5} A$ if and only if $\vdash_{5S} A$.

Proof. It is enough to show that $\vdash_{5S} \Diamond A \rightarrow \Box\Diamond A$, $\vdash_{S5} \Box A \rightarrow \Box\Box A$, and $\vdash_{S5} A \rightarrow \Box\Diamond A$.

$\vdash_{5S} \Diamond A \rightarrow \Box\Diamond A$: We have $\vdash_{5S} \Diamond A \rightarrow \Box\Diamond\Diamond A$ (because $\vdash_{5S} B \rightarrow \Box\Diamond B$, for all sentences B). We also have

$\vdash_{S5} \Diamond\Diamond A \to \Diamond A$ (because $\vdash_{S5} \Box B \to \Box\Box B$, for all B). By normality, $\vdash_{S5} \Box\Diamond\Diamond A \to \Box\Diamond A$, and so $\vdash_{S5} \Diamond A \to \Box\Diamond A$.

$\vdash_{S5} \Box A \to \Box\Box A$: Since $S5$ extends T, we have $\vdash_{S5} \Box A \to \Diamond\Box A$, and also $\vdash_{S5} \Diamond\Box A \to \Box\Diamond\Box A$ (because $\vdash_{S5} \Diamond B \to \Box\Diamond B$), and therefore

(*) $\quad \vdash_{S5} \Box A \to \Box\Diamond\Box A.$

Now, $\vdash_{S5} \Diamond -A \to \Box\Diamond -A$, and so, contraposing, we have $\vdash_{S5} \Diamond\Box A \to \Box A$, and then by normality, $\vdash_{S5} \Box\Diamond\Box A \to \Box\Box A$. By (*), $\vdash_{S5} \Box A \to \Box\Box A$.

$\vdash_{S5} A \to \Box\Diamond A$: This immediately follows from $\vdash_{S5} A \to \Diamond A$ and $\vdash_{S5} \Diamond A \to \Box\Diamond A$. $\dashv$

We shall now show that $\Box p \to p$ is *not* a theorem of G, and, therefore, that G is consistent. We do this by associating with each modal sentence A a sentence A^* and showing that if A is a theorem of G, then A^* is a tautology, and showing that $(\Box p \to p)^*$ is not a tautology. Let * be defined by: $\bot^* = \bot$; $p^* = p$ (for all sentence letters p); $(A \to B)^* = (A^* \to B^*)$; and $(\Box A)^* = \top$. (Then A^* is the result of taking $\Box$ to be a *verum* operator in A.) If A is a tautology, then so is A^*; if A is a distribution axiom, then $A^* = \top \to (\top \to \top)$, a tautology; for every A, $(\Box(\Box A \to A) \to \Box A)^* = (\top \to \top)$, a tautology; if A^* and $(A \to B)^*$, $= (A^* \to B^*)$, are tautologies, then so is B^*; and $(\Box A)^* = \top$, is certainly a tautology if A^* is a tautology. Thus if A is a sentence that occurs in a proof, that is, if A is a theorem, then A^* is a tautology. But $(\Box p \to p)^* = \top \to p$, which is equivalent to p and not a tautology. It follows that $\Box p \to p$ is not a theorem of G and hence that it is not a theorem of K either.

On the other hand, $\Box(\Box p \to p) \to \Box p$ is not a theorem of $S5$ and hence not a theorem of B, $S4$, T, $K4$, or K. For let $_*$ be defined by: $\bot_* = \bot$; $p_* = p$; $(A \to B)_* = (A_* \to B_*)$; and $(\Box A)_* = A_*$. (A_* is just the result of deleting all $\Box$s from A.) Then $(\Box(A \to A')$

$\rightarrow (\Box A \rightarrow \Box A'))_* = ((A_* \rightarrow A_*') \rightarrow (A_* \rightarrow A_*'))$, a tautology; $(\Box A \rightarrow A)_* = (A_* \rightarrow A_*)$, a tautology; and $(\Diamond A \rightarrow \Box\Diamond A)_* = (--A_* \rightarrow --A_*)$, a tautology. Thus, if A is a theorem of $S5$, then A_* is a tautology; but $(\Box(\Box p \rightarrow p) \rightarrow \Box p)_* = ((p \rightarrow p) \rightarrow p)$, which is not a tautology.

G and T are therefore consistent normal systems of modal logic, but there is no consistent normal system that extends both of G and T.

A remarkable fact about G, the proof of which was discovered independently by Kripke, de Jongh, and Sambin, is that $\Box p \rightarrow \Box\Box p$ is a theorem of G. (If $\Box p \rightarrow \Box\Box p$ had not been a theorem of G, we should have been interested in the smallest normal proper extension of G in which it was a theorem!)

Theorem 9

$\vdash_G \Box p \rightarrow \Box\Box p$.

Proof. $p \rightarrow (\Box p \,\&\, \Box\Box p \rightarrow p \,\&\, \Box p)$ is a tautology and, therefore, $\vdash_G p \rightarrow (\Box p \,\&\, \Box\Box p \rightarrow p \,\&\, \Box p)$. By Theorem 1, setting $A = p$ and $B = \Box p$, we have

(*) $\vdash_G \Box(p \,\&\, \Box p) \leftrightarrow (\Box p \,\&\, \Box\Box p)$. Therefore,
$\vdash_G p \rightarrow (\Box(p \,\&\, \Box p) \rightarrow p \,\&\, \Box p)$. By normality,
$\vdash_G \Box p \rightarrow \Box(\Box(p \,\&\, \Box p) \rightarrow p \,\&\, \Box p)$. But since
$\vdash_G \Box(\Box(p \,\&\, \Box p) \rightarrow p \,\&\, \Box p) \rightarrow \Box(p \,\&\, \Box p)$

– substitute $(p \,\&\, \Box p)$ for p in $\Box(\Box p \rightarrow p) \rightarrow \Box p$ – we have

$\vdash_G \Box p \rightarrow \Box(p \,\&\, \Box p)$, and therefore by (*),
$\vdash_G \Box p \rightarrow \Box p \,\&\, \Box\Box p$, and so
$\vdash_G \Box p \rightarrow \Box\Box p$. ⊣

Since G is normal, every theorem of $K4$ is thus a theorem of G. Theorem 5 thus holds when '$K4$' is replaced by 'G'.

Theorem 10

(a) $\vdash_G \Box(\Box A \rightarrow A) \leftrightarrow \Box A$.
(b) $\vdash_G -\Box A \leftrightarrow \Diamond(-A \mathbin{\&} \Box A)$.
(c) $\vdash_G \Diamond B \leftrightarrow \Diamond(B \mathbin{\&} \Box -B)$.
(d) $\vdash_G \Box A \leftrightarrow \Box(A \mathbin{\&} \Box A)$.
(e) $\vdash_G \Box A \rightarrow (\Diamond B \leftrightarrow \Diamond(B \mathbin{\&} A \mathbin{\&} \Box A))$.

Proof. (a) holds by normality; (b) comes from (a) by taking negations. For (c), substitute $-B$ for A in (b). (d) follows from Theorem 9; and (e) follows from (d) and the corollary to Theorem 3. ⊣

G is of interest because we can find out from it facts about PA, ZF, and other formal theories.

Theorem 11
$\vdash_G \Box\bot \leftrightarrow \Box\Diamond p$.

Proof. $\vdash_K \bot \rightarrow \Diamond p$, and so $\vdash_K \Box\bot \rightarrow \Box\Diamond p$. And since $\vdash_K p \rightarrow \top$, $\vdash_K \Diamond p \rightarrow \Diamond\top$, and $\vdash_K \Box\Diamond p \rightarrow \Box\Diamond\top$. But since $\vdash_K \Diamond\top \leftrightarrow -\Box\bot$, $\vdash_K \Diamond\top \leftrightarrow (\Box\bot \rightarrow \bot)$ and, therefore, $\vdash_K \Box\Diamond\top \leftrightarrow \Box(\Box\bot \rightarrow \bot)$. Since G extends K and $\vdash_G \Box(\Box\bot \rightarrow \bot) \rightarrow \Box\bot$, $\vdash_G \Box\Diamond p \leftrightarrow \Box\bot$ ⊣

In Chapter 3 we shall see how Theorem 11 can be regarded as telling us that PA asserts of each sentence S of PA that PA is inconsistent if and only if it is provable (in PA) that S is consistent (with PA). And in that chapter we shall also see how the fact that $\vdash_G \Box\Diamond\top \rightarrow \Box\bot$ can be regarded as telling us that (the contrapositive of) the second incompleteness theorem is a theorem of PA.

Since both sides of the biconditional in Theorem 7 are substitution instances of $\Box\Diamond p$, it follows from Theorem 11 that $\vdash_G \Box\Diamond A \leftrightarrow \Box\Diamond\Box\Diamond A$.

Our proof that $\Box p \rightarrow p$ is not a theorem of G cannot be used to show that $p \rightarrow \Box\Diamond p$ and $\Diamond p \rightarrow \Box\Diamond p$ are not theorems of G. But in Chapter 3 we shall see that $\Box\bot$ is not a theorem of G. It follows from Theorem 11 that neither $\top \to \Box\Diamond\top$ nor $\Diamond\top \to \Box\Diamond\top$ is a theorem of G, since both are equivalent in G to $\Box\bot$, and, hence, that $p \to \Box\Diamond p$ and $\Diamond p \to \Box\Diamond p$ are not theorems either.

Theorem 12

$\vdash_G -\Box\bot \leftrightarrow -\Box-\Box\bot$; $\vdash_G \Box(p \leftrightarrow -\Box p) \to \Box(p \leftrightarrow -\Box\bot)$.

Proof. That $\vdash_G -\Box\bot \leftrightarrow -\Box-\Box\bot$ follows from Theorem 11 by substituting $\top$ for p and taking negations. And since $\vdash_G (p \leftrightarrow -\Box p) \to (p \to -\Box p)$, $\vdash_G \Box(p \leftrightarrow -\Box p) \to (\Box p \to \Box-\Box p)$. By substituting $-p$ for p in Theorem 11, we have

$\vdash_G \Box-\Box p \leftrightarrow \Box\bot$, and so
$\vdash_G \Box(p \leftrightarrow -\Box p) \to (\Box p \to \Box\bot)$.

But since $\vdash_G \bot \to p$, $\vdash_G \Box\bot \to \Box p$, and so

$\vdash_G \Box(p \leftrightarrow -\Box p) \to (\Box p \leftrightarrow \Box\bot)$, whence by normality,

$\vdash_G \Box\Box(p \leftrightarrow -\Box p) \to \Box(\Box p \leftrightarrow \Box\bot)$. But since G extends $K4$,

$\vdash_G \Box(p \leftrightarrow -\Box p) \to \Box\Box(p \leftrightarrow -\Box p)$, and so
$\vdash_G \Box(p \leftrightarrow -\Box p) \to \Box(\Box p \leftrightarrow \Box\bot)$.

And since $p \leftrightarrow -\Box\bot$ is a truth-functional consequence of $p \leftrightarrow -\Box p$ and $\Box p \leftrightarrow \Box\bot$, we have

$\vdash_G \Box(p \leftrightarrow -\Box p) \,\&\, \Box(\Box p \leftrightarrow \Box\bot)$
$\to \Box(p \leftrightarrow -\Box\bot)$, and therefore
$\vdash_G \Box(p \leftrightarrow -\Box p) \to \Box(p \leftrightarrow -\Box\bot)$. ⊣

The first part of Theorem 12 tells us that PA asserts that PA is consistent if and only if the consistency of PA is not provable in PA; the second part tells us that PA asserts that if S is a "Gödel sentence," that is, if S is equivalent in PA to the assertion that S is unprovable, then S is equivalent to the assertion that PA is inconsistent. Many other facts about PA, including some that are new and interesting, can be learned from a study of G.

Exercise

Define $\Box^n \bot$ by $\Box^0\bot = \bot$; $\Box^{n+1} \bot = \Box\Box^n \bot$. Show that for no n, $\vdash_G \Box^n \bot$. [*Hint:* for each n, define A^n by $\bot^n = \bot$; $p^n = p$; $(A \rightarrow B)^n = (A^n \rightarrow B^n)$; and $(\Box A)^n = \top \mathbin{\&} A^0 \mathbin{\&} A^1 \mathbin{\&} \cdots \mathbin{\&} A^{n-1}$. Show that for each n, A^n is a tautology if $\vdash_G A$, and $(\Box^n \bot)^n$ is equivalent to $\bot$ in the propositional calculus.]

2

Peano Arithmetic

The aim of the present chapter is to review the concepts mentioned in, and sketch the proofs of, five important theorems about Bew(x), the standard "provability" or "theoremhood" predicate of Peano Arithmetic (PA):

(For all sentences S, S' of Peano Arithmetic)

(i) If $\vdash_{\mathrm{PA}} S$, then $\vdash_{\mathrm{PA}} \mathrm{Bew}(\ulcorner S \urcorner)$,

(ii) $\vdash_{\mathrm{PA}} \mathrm{Bew}(\ulcorner (S \to S') \urcorner) \to (\mathrm{Bew}(\ulcorner S \urcorner) \to \mathrm{Bew}(\ulcorner S' \urcorner))$,

(iii) $\vdash_{\mathrm{PA}} \mathrm{Bew}(\ulcorner S \urcorner) \to \mathrm{Bew}(\ulcorner \mathrm{Bew}(\ulcorner S \urcorner) \urcorner)$,

(iv) $\mathrm{Bew}(\ulcorner S \urcorner)$ is a Σ_1-sentence, and

(v) if S is a Σ_1-sentence, then $\vdash_{\mathrm{PA}} S \to \mathrm{Bew}(\ulcorner S \urcorner)$.

$\ulcorner S \urcorner$ is the numeral in PA for the Gödel number of sentence S, that is, **0** followed by that number of occurrences of the successor sign $'$. $\mathrm{Bew}(\ulcorner S \urcorner)$ is thus the result of substituting $\ulcorner S \urcorner$ for all free occurrences of the variable x in Bew(x), and (iii) thus obviously follows from (iv) and (v). [$\mathrm{Bew}(\ulcorner S \urcorner)$ may be regarded as a sentence asserting that S is a theorem of PA.] A Σ_1-sentence, roughly speaking, is a sentence asserting that a certain primitive recursive set is nonempty. (A precise definition is given below.)

(v) and the notion of a Σ_1-sentence are used only in Chapter 12, in which Solovay's completeness theorems are proved. By the time we reach Chapter 12, borne by $K4$ and G, we shall have established a number of striking results about the notions of provability, consistency, relative consistency, and diagonal sentences (fixed points). The reader who does not like highly incomplete and (apparently) irremediably messy proofs of syntactic facts, may wish to skim over the rest of this chapter and take the truth of (i), (ii), and (iii) for granted.

Peano Arithmetic (classical first-order formal arithmetic with induction, or, simply, arithmetic) is a theory that is formalized in the classical first-order predicate calculus with identity. The language of PA contains just the individual constant **0**, the 1-ary function symbol $'$, and two 2-ary function symbols $+$ and $\cdot$. (Since zero can be proved in PA to be the unique number k such that $j + k = j$ for all j, and the successor of i can be proved to be the unique k such that $j \cdot k = j \cdot i + j$ for all j, **0** and $'$ are really inessential, but it simplifies matters somewhat and is traditional for them to be taken as primitives.) The theorems of PA are the logical consequences of the universal closures of the recursion axioms for successor, sum, and product, which are the six formulas,

(1) $\mathbf{0} \neq x'$,

(2) $x' = y' \to x = y$,

(3) $x + \mathbf{0} = x$,

(4) $x + y' = (x + y)'$,

(5) $x \cdot \mathbf{0} = \mathbf{0}$, and

(6) $x \cdot y' = x \cdot y + x$,

and the induction axioms, which are the instances of the induction schema

$$(A_x(\mathbf{0}) \ \& \ \forall x(A \to A_x(x')) \to \forall x A.$$

The variables of PA, in their "canonical" order, are $x_1, x_2, \ldots$. A *predicate* is a formula in which at most one variable occurs free. We shall say that a formula F *implies* a formula G if $\vdash_{\mathrm{PA}} F \to G$ and that F *is equivalent to* G if $\vdash_{\mathrm{PA}} F \leftrightarrow G$.

Sentences of PA will be called true or false when they

are true or false in the standard model of the language of PA, whose domain is the set of all natural numbers, and which assigns zero, the successor function, the addition function, and the multiplication function to **0**, ′, +, and •, respectively.

We shall suppose that the reader is already somewhat familiar with some standard treatment of Gödel's incompleteness theorems, that (s)he knows what Gödel numbering is, that (s)he has some rough idea of what primitive recursive functions and relations are (their definition is reviewed below), and that (s)he is willing to take for granted certain facts about primitive recursiveness, such as the primitive recursiveness of the relation $\{\langle m,n\rangle | m$ is the Gödel number of a proof in PA of the formula with Gödel number $n\}$. We shall also suppose that (s)he is willing to accept certain claims to the effect that certain formulas can be proved in PA.

We shall assume that we are discussing some version of PA in which we have the definite description operator ι available (by means of which terms can be formed from formulas). Under the standard conventions, every formula in which ι occurs is equivalent to one in which it does not occur; we shall suppose that ι-terms denote zero when they do not have their ordinary denotation. We shall also assume that in our version of PA there are finitely many schemata (of a standard sort), the closures of whose instances, together with those of the induction schema and (1)–(6), form the set of axioms; in which modus ponens and generalization are the rules of inference; and in which $\perp$ and $\rightarrow$ are two of the primitive logical symbols. $-\text{Bew}(\ulcorner\perp\urcorner)$ thus expresses the consistency of PA.

A proof in PA is, as usual, a sequence of formulas, each of which is either an axiom or a consequence, by modus ponens or generalization, of earlier formulas in the sequence. A proof is a proof *of* its last formula, and a theorem is a formula of which there is a proof.

Numerals. The numeral for the number n is the result of attaching $'$ to $\mathbf{0}$ n times. Thus the numeral for 3 is $\mathbf{0}'''$, the numeral for 1 is $\mathbf{0}'$, and the numeral for 0 is $\mathbf{0}$. The numeral for n is denoted by '$\mathbf{n}$'; thus $\mathbf{3}$ is $\mathbf{0}'''$. $\ulcorner F \urcorner$ is the numeral for the Gödel number of F.

We shall abbreviate '$x_1, \ldots, x_n$', '$i_1, \ldots, i_n$', and '$\mathbf{i}_1, \ldots, \mathbf{i}_n$' by '$\mathsf{x}$', '$\mathsf{i}$', and '$\mathbf{i}$'.

Representing. A term $t(\mathsf{x})$ is said to *represent* an n-ary function f (in PA) if for all i, i_{n+1}, if $f(\mathsf{i}) = i_{n+1}$, then $\vdash_{\mathrm{PA}} t(\mathbf{i}) = \mathbf{i}_{n+1}$.

Thus, for example, $x_1 \cdot x_2$ represents the multiplication function.

Primitive recursive functions. The *primitive recursive functions* are the members of the smallest class that contains the zero, successor, and identity functions and that contains all functions that come from members by compostion and primitive recursion. The *zero function* is the function whose value is zero for every natural number. The *successor function* is the function whose value is $i + 1$ for every natural number i. For every m, k, $1 \leqslant k \leqslant m$, there is an *identity* (or *projection*) *function* $\mathrm{id}_k{}^m$: $\mathrm{id}_k{}^m\,(i_1, \ldots, i_m) = i_k$; if f is an m-ary function and $g_1, \ldots, g_m$ are n-ary functions, then h *comes from f and $g_1, \ldots, g_m$ by composition* if for all i, $h(\mathsf{i}) = f(g_1(\mathsf{i}), \ldots, g_m(\mathsf{i}))$; and if f is an n-ary function and g an $(n + 2)$-ary function, then h *comes from f and g by primitive recursion* if $h(\mathsf{i},0) = f(\mathsf{i})$ and $h(\mathsf{i},k + 1) = g(\mathsf{i},k,h(\mathsf{i},k))$ (all i,k). [If h comes from the n-ary f and the $(n + 2)$-ary g by primitive recursion, then h is $(n + 1)$-ary.] A *primitive recursive relation* is one whose characteristic function is primitive recursive. (In the present treatment, 0 plays the role of truth.) We shall identify sets with 1-ary relations and relations with their characteristic functions.

The beta-function. In "On formally undecidable propositions . . . ," Gödel showed that a certain function β, whose value for natural numbers a, b, k is the remainder on dividing a by $1 + (k + 1)b$, possesses the following property:

> For any natural number m and any finite sequence $h_0, h_1, \ldots, h_m$ of natural numbers, there exist natural numbers a, b such that for all $k \leqslant m$, $\beta(a, b, k) = h_k$.

The proof of this fact uses a well-known elementary result from number theory, known as the Chinese Remainder Theorem; the details are easily available elsewhere (e.g., in Section 48 of Kleene's *Introduction to Metamathematics*[1]) and we shall omit them.

(β is a primitive recursive function.)

Suppose that $s(\mathbf{x})$ represents f, $t(\mathbf{x}, x_{n+1}, x_{n+2})$ represents g, and h comes from f and g by primitive recursion. We want to see how to use β to construct a term $u(\mathbf{x}, x_{n+1})$ that represents h from the terms $s(\mathbf{x})$ and $t(\mathbf{x}, x_{n+1}, x_{n+2})$.

Since h comes from f and g by primitive recursion, $h(\mathbf{i}, 0) = f(\mathbf{i})$ and $h(\mathbf{i}, k + 1) = g(\mathbf{i}, k, h(\mathbf{i}, k))$. Thus $h(\mathbf{i}, m) = j$ if and only if there is a finite sequence $h_0, h_1, \ldots, h_m$ such that $h_0 = f(\mathbf{i})$, $h_{k+1} = g(\mathbf{i}, k, h_k)$ for all $k < m$, and $h_m = j$. The fundamental property of β enables us to replace the "quantification over" finite sequences of natural numbers in the previous assertion by a (double) quantification over natural numbers: Thus $h(\mathbf{i}, m) = j$ if and only if there exist natural numbers a, b such that $\beta(a, b, 0) = f(\mathbf{i})$, $\beta(a, b, k + 1) = g(i, k, \beta(a, b, k))$ for all $k < m$, and $\beta(a, b, m) = j$.

We define $x < y$ to be the formula $\exists z\, z' + x = y$.

And noting that $\beta(a, b, k)$ is the unique number r less than $1 + (k + 1)b$ such that for some i, $a = i(1 + (k + 1)b) + r$, we define $\mathrm{Beta}(w_1, w_2, z)$ to be the term

$$\iota y(y < (\mathbf{1} + (z + \mathbf{1}) \cdot w_2)$$
$$\& \exists x(w_1 = x \cdot (\mathbf{1} + (z + \mathbf{1}) \cdot w_2) + y)).$$

We can now construct our desired term $u(\mathbf{x}, x_{n+1})$. We let $u(\mathbf{x}, x_{n+1})$ be the term

$$\iota y \exists w_1 \exists w_2(\text{Beta}(w_1, w_2, \mathbf{0}) = s(\mathbf{x}) \;\&\; \forall z(z < x_{n+1}$$
$$\rightarrow \text{Beta}(w_1, w_2, z') = t(\mathbf{x}, z, \text{Beta}(w_1, w_2, z)))$$
$$\& \text{Beta}(w_1, w_2, x_{n+1}) = y).$$

We shall say that $u(\mathbf{x}, x_{n+1})$ *comes from* $s(\mathbf{x})$ *and* $t(\mathbf{x}, x_{n+1}, x_{n+2})$ *by term recursion*. We omit the (standard) proof that $u(\mathbf{x}, x_{n+1})$ represents h.

(A quite different but equally serviceable beta-function is described in Chapter 14 of *Computability and Logic.*[2])

Primitive recursive terms. The *primitive recursive terms* have an inductive definition that parallels that of the primitive recursive functions: They are the members of the smallest class that contains the zero, successor, and identity terms and that contains all terms that come from members by term composition and term recursion.

The zero term is the term $\iota y(x_1 = x_1 \;\&\; y = \mathbf{0})$. The zero term represents the zero function.

The successor term is the term $x_1{}'$. The successor term represents the successor function.

The identity term $\text{Id}_k{}^m$ $(1 \leqslant k \leqslant m)$ is the term $\iota y(x_1 = x_1 \;\&\; \cdots \;\&\; x_m = x_m \;\&\; y = x_k)$. $\text{Id}_k{}^m$ represents $\text{id}_k{}^m$

The term that *comes from* $s(x_1, \ldots, x_m)$ *and* $t_1(\mathbf{x}), \ldots, t_m(\mathbf{x})$ *by term composition* is $s(t_1(\mathbf{x}), \ldots, t_m(\mathbf{x}))$. If $s(x_1, \ldots, x_m)$ represents f, $t_1(\mathbf{x})$ represents $g_1, \ldots, t_m(\mathbf{x})$ represents g_m, and h comes from f and $g_1, \ldots, g_m$ by composition, then $s(t_1(\mathbf{x}), \ldots, t_m(\mathbf{x}))$ represents h.

Every primitive recursive term represents one primitive recursive function; but every primitive recursive function is represented by infinitely many primitive recursive terms. A primitive recursive term is said to be *for* the function it represents.

In "On formally undecidable propositions. . . " there is a list of forty-six functions and relations, of which all but the last, the set of Gödel numbers of theorems of Gödel's system P, are primitive recursive. (Gödel used the term 'recursive', which has subsequently come to mean 'general recursive'; the general recursive functions form a strictly wider class than the primitive recursive functions.) Item 45 on the list is the "proof-of" relation (for P), and each function or relation on the list either is obviously primitive recursive or can be shown to be primitive recursive from the assumption that earlier items on the list are primitive recursive. Several short-cuts, justified in advance, were used to abbreviate the list to forty-six items.

It will be useful to introduce, informally, and for the nonce, the notion of a prim rec list. (Gödel's list can be thought of as an abbreviated prim rec list.) A *prim rec list* is a finite list of primitive recursive functions, indexed by consecutive positive integers, each item on which is (correctly) annotated as being the zero function, the successor function, or id_k^m, or as coming from earlier functions by one of the operations of composition and primitive recursion. The annotation must specify the index numbers of the earlier functions and specify the operation (and in the case of composition, must also specify which function plays the role of f and which functions play the roles of $g_1, \ldots, g_m$).

For each prim rec list, we associate a primitive recursive term with each index number of the list by using the annotations in the obvious way: Suppose that terms have been associated with all index numbers $<n$. If item n is annotated as being the zero function, the successor function, or id_k^m, associate with n the zero term, the successor term, or Id_k^m, respectively. If item n is annotated as having come from items i and $j_1, \ldots, j_m$ by composition, associate with n the term that comes from those associated with i and $j_1, \ldots, j_m$ by term compo-

sition; and do the obvious analogous thing in the case of primitive recursion. Then each term associated with an index number of the list represents the function with that number.

We shall suppose that we possess a prim rec list whose last item is the "proof-of" relation for PA, that is, the relation $\{\langle m,n\rangle | m$ is the Gödel number of a proof in PA of the formula with Gödel number $n\}$. We shall also suppose that among the items on the list are the set of Gödel numbers of axioms of PA, the relations $\{\langle i,j,k\rangle | i$ is the Gödel number of an immediate consequence by modus ponens of the formulas with Gödel numbers j and $k\}$, $\{\langle i,j\rangle | i$ is the Gödel number of an immediate consequence by generalization of the formula with Gödel number $j\}$, and $\{\langle i,j,k\rangle | i$ is the Gödel number of the jth component of the sequence with Gödel number $k\}$.

Finally, we shall suppose that an examination of the details of the annotations would show that the proof-of relation was obtained *in the usual way* from the axiom set, the two immediate consequence relations, and the sequence–component relation. If the proof-of relation is obtained in some unusual way, we have no guarantee that the predicate Bew(x) that we shall construct from our prim rec list will satisfy (i)–(v).

Let Prf(x_1,x_2) be the term associated with the last index number of our list. Then Pf(x_1,x_2) is to be the formula Prf(x_1,x_2) $= \mathbf{0}$, and Bew(x) is to be the formula $\exists y$Pf(y,x).

That (i) holds follows from the fact that each term associated with an index number of our list represents the function indexed by that number: Let prf be (the characteristic function of) the proof-of relation. Then Prf(x_1,x_2) represents prf. Suppose that $\vdash_{\mathrm{PA}} S$. Then there is a proof P of S in PA. Let m be the Gödel number of P and n be the Gödel number of S. Then prf(m,n) $= 0$. Since Prf(x_1,x_2) represents prf and $\mathbf{n} = \ulcorner S \urcorner$,

$\vdash_{PA}$ Prf($\mathbf{m}$,$\ulcorner S \urcorner$) = **0**, that is, $\vdash_{PA}$ Pf($\mathbf{m}$,$\ulcorner S \urcorner$), and so $\vdash_{PA}$ $\exists y$ Pf(y,$\ulcorner S \urcorner$), that is, $\vdash_{PA}$ Bew($\ulcorner S \urcorner$).

We must now explain a piece of notation: 'Bew[F]'.

Let sub be the primitive recursive function such that sub(i, j, k) = the Gödel number of the result of substituting $\mathbf{i}$ for all free occurrences of the variable x_j in the formula with Gödel number k. sub(i, j, k) = 0 if k is not the Gödel number of a formula; sub(i, j, k) = k if x_j does not occur free in the formula with Gödel number k. For example, if k is the Gödel number of the formula $x_4 = x_1$, then sub(3, 4, k) = the Gödel number of the formula $\mathbf{3} = x_1$.

Let Sub(x_1,x_2,x_3) be a primitive recursive term for sub. Then, to continue the example, $\vdash_{PA}$ Sub(**3**,**4**,$\ulcorner x_4 = x_1 \urcorner$) = $\ulcorner \mathbf{3} = x_1 \urcorner$.

Suppose now that F is a formula of PA with precisely the m variables $x_{k_1}, x_{k_2}, \ldots, x_{k_m}$ free. We define Bew[F] to be the formula

$$\text{Bew}(\text{Sub}(x_{k_1}, \mathbf{k}_1, \text{Sub}(x_{k_2}, \mathbf{k}_2, \ldots, \text{Sub}(x_{k_m}, \mathbf{k}_m, \ulcorner F \urcorner) \ldots))).$$

If F has no free variables, that is, if F is a sentence, then Bew[F] is to be the sentence Bew($\ulcorner F \urcorner$).

Thus Bew[F] has the same free variables, namely, x_{k_1}, . . . , x_{k_m} (or none), as F. Bew[F] is true under the assignment of i_1 to x_{k_1}, . . . , and i_m to x_{k_m} if and only if $F_{x_{k_1} \cdots x_{k_m}}(\mathbf{i}_1, \ldots, \mathbf{i}_m)$, which is the result of substituting $\mathbf{i}_1$ for x_{k_1}, . . . , and $\mathbf{i}_m$ for x_{k_m} in F, is provable in PA.

We can now state a fundamental result about Bew(x) and primitive recursive terms: If $t(\mathbf{x})$ is a primitive recursive term, then $\vdash_{PA}$ $t(\mathbf{x}) = x_{n+1} \rightarrow$ Bew[$t(\mathbf{x}) = x_{n+1}$].

The proof of this result is a formalization of the proof that each term associated with an index number of a prim rec list represents the function indexed by that number. The simplest case, corresponding to the term x_1', which is the proof that $\vdash_{PA}$ $x_1' = x_2 \rightarrow$ Bew[$x_1' = x_2$], is a formalization of the following argument: If $i_1' = i_2$, then the result of attaching a successor sign to $\mathbf{i}_1$ *is* $\mathbf{i}_2$, and hence the result of substituting $\mathbf{i}_1$ for x_1 and $\mathbf{i}_2$ for x_2 in $x_1' = x_2$ is $\mathbf{i}_2 = \mathbf{i}_2$, and is therefore a theorem of PA. (A detailed treatment of this result, for a system akin to PA, is found in Hilbert–Bernays, *Grundlagen der Mathematik.*[3] A very useful discussion can also be found in Shoenfield, *Mathematical Logic.*[4])

With the notation 'Bew[F]' and the result about primitive recursive terms and Bew(x) in hand, we can dispose of (ii)–(v) fairly quickly.

We first want to see that (A) if $\vdash_{PA} F$, then $\vdash_{PA}$ Bew[F]. Let S be the universal closure of F. It suffices to show that $\vdash_{PA}$ Bew($\ulcorner S \urcorner$) $\rightarrow$ Bew[F], for then, if $\vdash_{PA} F$, then by generalization, $\vdash_{PA} S$, whence by (i) $\vdash_{PA}$ Bew($\ulcorner S \urcorner$), and so $\vdash_{PA}$ Bew[F]. But a formalization in PA of the following argument shows that $\vdash_{PA}$ Bew($\ulcorner S \urcorner$) $\rightarrow$ Bew[F]: Since S is the universal closure of F, we can construct a proof of $F_{x_{k_1}\cdots x_{k_m}}(\mathbf{i}_1, \ldots, \mathbf{i}_m)$ from a proof of S; thus, if S is provable, so is $F_{x_{k_i}\cdots x_{k_m}}(\mathbf{i}_1, \ldots, \mathbf{i}_m)$.

The formalization of the following argument shows that (B) $\vdash_{PA}$ Bew[$(F \rightarrow F')$] $\rightarrow$ (Bew[F] $\rightarrow$ Bew[F']): Since modus ponens is a rule of PA, from proofs of the results of a substitution of certain numerals for certain variables in both $(F \rightarrow F')$ and F, we can construct a proof of the result of that substitution of those numerals for those variables in F'.

(ii) is the special case of (B) in which F and F' are sentences.

(C) if F implies G, then Bew[F] implies Bew[G]: For by (B), $\vdash_{PA}$ Bew[$(F \rightarrow G)$] $\rightarrow$ (Bew[F] $\rightarrow$ Bew[G]), and, thus, if F implies G, that is, if $\vdash_{PA} F \rightarrow G$, then by (A), $\vdash_{PA}$ Bew[$(F \rightarrow G)$], and so $\vdash_{PA}$ Bew[F] $\rightarrow$ Bew[G].

(D) If F is equivalent to G, then Bew[F] is equivalent to Bew[G].

Primitive recursive formulas and Σ_1-formulas. A *primitive recursive formula in the strong sense* is a formula $t(\mathbf{x}) = \mathbf{0}$, where $t(\mathbf{x})$ is a primitive recursive term.

A *primitive recursive formula* is one that is equivalent to a primitive recursive formula in the strong sense. Thus F is a primitive recursive formula if $\vdash_{PA} F \leftrightarrow t(\mathbf{x}) = \mathbf{0}$, for some primitive recursive term $t(\mathbf{x})$. The primitive recursive formulas are closed under operations analogous to those under which the primitive recursive relations are closed, for example, substitution of numerals and primitive recursive terms, changes and identifications of variables, truth-functional operations, and *bounded* quantification. $\mathrm{Pf}(y,x)$ is a primitive recursive formula.

A Σ_1*-formula in the strong sense* is an existential quantification with respect to a single variable of a primitive recursive formula in the strong sense.

A Σ_1*-formula* is one that is equivalent to a Σ_1-formula in the strong sense. The Σ_1-formulas are closed under substitution of numerals and primitive recursive terms, changes and identifications of variables, conjunction, disjunction, bounded quantification, and existential quantification (this last because of the existence of primitive recursive pairing functions), but not under unbounded universal quantification, and hence not under negation. $\mathrm{Bew}(x)$ is a Σ_1-formula. Indeed, if F is any formula, then $\mathrm{Bew}[F]$ is a Σ_1-formula, since $\mathrm{Sub}(x_1,x_2,x_3)$ is a primitive recursive term.

A Σ_1*-sentence* is just a Σ_1-formula that is a sentence. Thus (iv) holds.

We shall prove (v) and hence (iii) by proving a result about PA, the so-called provable Σ_1-completeness of PA, of which (v) is the special case in which F is a sentence:

If F is a Σ_1-formula, then $\vdash_{PA} F \rightarrow \mathrm{Bew}[F]$.

Suppose that F is a Σ_1-formula. Then there exists a primitive recursive term $t(\mathbf{x},x_{n+1})$ such that $\vdash_{PA} F \leftrightarrow \exists x_{n+1}\, t(\mathbf{x},x_{n+1}) = \mathbf{0}$. By (D) it suffices to show that

$$\vdash_{PA} \exists x_{n+1}\, t(\mathbf{x},x_{n+1}) = \mathbf{0} \\ \rightarrow \mathrm{Bew}[\exists x_{n+1}\, t(\mathbf{x},x_{n+1}) = \mathbf{0}]$$

We know that

$$(*) \qquad \vdash_{PA} t(\mathbf{x}, x_{n+1}) = x_{n+2} \\ \rightarrow \mathrm{Bew}[t(\mathbf{x}, x_{n+1}) = x_{n+2}].$$

Because

$$\vdash_{PA} \mathrm{Sub}(\mathbf{0},\mathbf{n+2}, \ulcorner t(\mathbf{x}, x_{n+1}) = x_{n+2}\urcorner) \\ = \ulcorner t(\mathbf{x}, x_{n+1}) = \mathbf{0}\urcorner,$$

it follows by substituting $\mathbf{0}$ for x_{n+2} in (*) that

$$\vdash_{PA} t(\mathbf{x}, x_{n+1}) = \mathbf{0} \rightarrow \mathrm{Bew}[t(\mathbf{x}, x_{n+1}) = \mathbf{0}].$$

But since $t(\mathbf{x}, x_{n+1}) = \mathbf{0}$ implies $\exists x_{n+1}\, t(\mathbf{x}, x_{n+1}) = \mathbf{0}$, by (C) we have that

$$\vdash_{PA} \mathrm{Bew}[t(\mathbf{x}, x_{n+1}) = \mathbf{0}] \\ \rightarrow \mathrm{Bew}[\exists x_{n+1}\, t(\mathbf{x}, x_{n+1}) = \mathbf{0}].$$

and thus that

$$\vdash_{PA} t(\mathbf{x}, x_{n+1}) = \mathbf{0} \rightarrow \mathrm{Bew}[\exists x_{n+1}\, t(\mathbf{x}, x_{n+1}) = \mathbf{0}].$$

And since x_{n+1} does not occur free in the consequent of this last formula, we have

$$\vdash_{PA} \exists x_{n+1}\, t(\mathbf{x}, x_{n+1}) = \mathbf{0} \\ \rightarrow \mathrm{Bew}[\exists x_{n+1}\, t(\mathbf{x}, x_{n+1}) = \mathbf{0}],$$

and therefore $\vdash_{PA} F \rightarrow \mathrm{Bew}[F]$.

3

The box as Bew

Our chief aim in this study is to investigate the effects of interpreting the box of modal logic to mean 'it is provable (in a certain formal theory) that . . . '. When modal logic is viewed this way, a question immediately comes to mind, to which the answer does not by any means appear to be evident, namely, Which principles of modal logic are correct when the box is interpreted in this way? The answer is *not* evident; near the end of this chapter we say what that answer is, and in Chapter 12, in which we prove Solovay's completeness theorems, we show that it is the answer.

In order to express our question in a mathematically precise way, we make two definitions.

A *realization* is a function that assigns to each sentence letter a sentence of the language of Peano Arithmetic.

The *translation* A^ϕ of a modal sentence A under a realization ϕ is defined inductively:

(1) $\bot^\phi = \bot$;
(2) $p^\phi = \phi(p)$ (for each sentence letter p);
(3) $(A \rightarrow B)^\phi = (A^\phi \rightarrow B^\phi)$;
(4) $(\Box A)^\phi = \mathrm{Bew}[A^\phi]$ $(= \mathrm{Bew}(\ulcorner A^\phi \urcorner))$.

We have taken $\bot$ and $\rightarrow$ to be among the primitive logical symbols of PA, and therefore the translation of any modal sentence under any realization is a sentence of the language of PA. Clauses (1) and (3) guarantee that the translation (under ϕ) of a truth-functional combination of sentences is that same truth-functional combination of the translations of those sentences. Clause (4) ensures that if the translation of A is S, then the translation of $\Box A$ is Bew[S], the result of substituting the numeral for the Gödel number of S for the free variable x in Bew(x), which is a sentence of Peano Arithmetic that may be regarded as expressing the assertion that S is provable.

If A is a letterless sentence, then $A^{\phi_1} = A^{\phi_2}$ for all realizations ϕ_1, ϕ_2, and thus 'A^ϕ' may be regarded as possessing a definite denotation even in the absence of any specification of ϕ, namely A^ϕ, where ϕ is any realization whatsoever.

Our original question, Which principles of modal logic are correct if the box is taken to mean 'it is provable that . . . '?, now gives way to two precisely formulated questions: Which modal sentences A are such that, for all realizations ϕ, A^ϕ is true (in the standard model)? and Which modal sentences A are such that, for all realizations ϕ, A^ϕ is provable (in PA)? Even though it is much more likely that the first reformulation expresses the intended meaning of the original question than that the second does, both are interesting questions, and we shall finally answer the first only after we have answered the second.

Recall from Chapter 1 the modal calculus $K4$. Its axioms are all tautologies, distribution axioms, and modal sentences of the form $\Box A \rightarrow \Box\Box A$, and its rules of inference are modus ponens and necessitation. We shall show below that if A is a theorem of G, then, for every realization ϕ, A^ϕ is a theorem of PA. In order to do this we first show that the same holds for the subsystem $K4$ of G.

Theorem 1
If $\vdash_{K4} A$, then, for every realization ϕ, $\vdash_{PA} A^{\phi}$.

Proof. If A is a tautological combination of modal sentences, then A^{ϕ} is a tautological combination of sentences of the language of PA, and therefore $\vdash_{PA} A^{\phi}$.

In Chapter 2 we saw that for every pair of sentences S, S' of the language of PA, $\vdash_{PA} \text{Bew}[(S \rightarrow S')] \rightarrow (\text{Bew}[S] \rightarrow \text{Bew}[S'])$. Thus, for every realization ϕ and every pair of modal sentences A, A', $\vdash_{PA} \text{Bew}[(A^{\phi} \rightarrow A'^{\phi})] \rightarrow (\text{Bew}[A^{\phi}] \rightarrow \text{Bew}[A'^{\phi}])$. Since $\text{Bew}[(A^{\phi} \rightarrow A'^{\phi})] \rightarrow (\text{Bew}[A^{\phi}] \rightarrow \text{Bew}[A'^{\phi}]) = (\Box(A \rightarrow A') \rightarrow (\Box A \rightarrow \Box A'))^{\phi}$, we have that, for every pair of modal sentences A, A', $\vdash_{PA} (\Box(A \rightarrow A') \rightarrow (\Box A \rightarrow \Box A'))^{\phi}$.

In Chapter 2 we also saw that for every sentence S of the language of PA, $\vdash_{PA} \text{Bew}[S] \rightarrow \text{Bew}[\text{Bew}[S]]$. Thus, for every realization ϕ and every modal sentence A, $\vdash_{PA} \text{Bew}[A^{\phi}] \rightarrow \text{Bew}[\text{Bew}[A^{\phi}]]$. Since $\text{Bew}[A^{\phi}] \rightarrow \text{Bew}[\text{Bew}[A^{\phi}]] = (\Box A \rightarrow \Box\Box A)^{\phi}$, we have that, for every modal sentence A, $\vdash_{PA} (\Box A \rightarrow \Box\Box A)^{\phi}$.

If $\vdash_{PA} A^{\phi}$ and $\vdash_{PA} (A \rightarrow B)^{\phi}$, then $\vdash_{PA} B^{\phi}$, since $(A \rightarrow B)^{\phi} = (A^{\phi} \rightarrow B^{\phi})$.

Lastly, if $\vdash_{PA} A^{\phi}$, then, as we also saw in Chapter 2, $\vdash_{PA} \text{Bew}[A^{\phi}]$, and thus $\vdash_{PA} (\Box A)^{\phi}$, since $\text{Bew}[A^{\phi}] = (\Box A)^{\phi}$.

It follows that if A is a theorem of $K4$, then A^{ϕ} is a theorem of PA. ⊣

We saw in Chapter 1 that $\Box(\Box p \rightarrow p) \rightarrow \Box p$, which is a theorem of G, is not a theorem of $S5$ and therefore not a theorem of $K4$. In order to see that every translation of every theorem of G is a theorem of PA, and thus that the converse of Theorem 1 is false, we shall prove a theorem about $K4$, prove the generalized diagonal lemma, and then make an easy application of the lemma.

Theorem 2
$\vdash_{K4} \Box(q \leftrightarrow (\Box q \rightarrow p)) \rightarrow (\Box(\Box p \rightarrow p) \rightarrow \Box p)$.

Proof

(1) $\vdash_{K4} \Box(q \leftrightarrow (\Box q \rightarrow p)) \rightarrow (\Box q \rightarrow \Box(\Box q \rightarrow p))$, by normality of *K*4.

(2) $\vdash_{K4} \Box(\Box q \rightarrow p) \rightarrow (\Box\Box q \rightarrow \Box p)$; (2) is a distribution axiom.

(3) $\vdash_{K4} \Box q \rightarrow \Box\Box q$.

(4) $\vdash_{K4} \Box(q \leftrightarrow (\Box q \rightarrow p)) \rightarrow (\Box q \rightarrow \Box p)$; (4) follows from (1), (3), and (2) by the propositional calculus.

(5) $\vdash_{K4} \Box\Box(q \leftrightarrow (\Box q \rightarrow p)) \rightarrow \Box(\Box q \rightarrow \Box p)$; (5) follows from (4) by normality.

(6) $\vdash_{K4} \Box(q \leftrightarrow (\Box q \rightarrow p)) \rightarrow \Box\Box(q \leftrightarrow (\Box q \rightarrow p))$; (6) is of the form $\Box A \rightarrow \Box\Box A$.

(7) $\vdash_{K4} \Box(\Box p \rightarrow p) \rightarrow (\Box(\Box q \rightarrow \Box p) \rightarrow \Box(\Box q \rightarrow p))$, by normality.

(8) $\vdash_{K4} \Box(q \leftrightarrow (\Box q \rightarrow p)) \rightarrow (\Box(\Box q \rightarrow p) \rightarrow \Box q)$, by normality.

(9) $\vdash_{K4} \Box(q \leftrightarrow (\Box q \rightarrow p)) \rightarrow (\Box(\Box p \rightarrow p) \rightarrow \Box p)$; (9) follows from (6), (5), (7), (8), and (4) by the propositional calculus. $\dashv$

The proof of Theorem 2 is a formalization in *K*4 of a part of the argument contained in Löb's proof that if $\vdash_{PA}$ Bew[*S*] $\rightarrow$ *S*, then $\vdash_{PA}$ *S*.[1]

Here is the generalized diagonal lemma.

The generalized diagonal lemma

Suppose that $P_1(y_1, \ldots, y_n, \mathbf{z}), \ldots, P_n(y_1, \ldots, y_n, \mathbf{z})$ are formulas of the language of PA in which all freely occurring variables are among $y_1, \ldots, y_n$, $\mathbf{z}$ ('$\mathbf{z}$' abbreviates '$z_1, \ldots, z_m$'). Then there exist formulas $S_1(\mathbf{z}), \ldots, S_n(\mathbf{z})$ of the language of PA in which all freely occurring variables are among $\mathbf{z}$, such that $\vdash_{PA} S_1(\mathbf{z}) \leftrightarrow P_1(\ulcorner S_1(\mathbf{z})\urcorner, \ldots, \ulcorner S_n(\mathbf{z})\urcorner, \mathbf{z}), \ldots$, and $\vdash_{PA} S_n(\mathbf{z}) \leftrightarrow P_n(\ulcorner S_1(\mathbf{z})\urcorner, \ldots, \ulcorner S_n(\mathbf{z})\urcorner, \mathbf{z})$.

Proof. Define a primitive recursive function su by su$(a, b_1, \ldots, b_n)$ = the Gödel number of the result of respectively substituting the numerals $\mathbf{b}_1, \ldots, \mathbf{b}_n$ for all free occurrences of $x_1, \ldots, x_n$ in the formula with Gödel number a.

Let $\mathrm{Su}(x_1, x_2, \ldots, x_{n+1})$ be a primitive recursive term for su.

For each i, $1 \leqslant i \leqslant n$, let k_i be the Gödel number of $P_i(\mathrm{Su}(x_1, x_1, \ldots, x_n), \ldots, \mathrm{Su}(x_n, x_1, \ldots, x_n), \mathbf{z})$, and let $S_1(\mathbf{z})$ be the formula $P_i(\mathrm{Su}(\mathbf{k}_1, \mathbf{k}_1, \ldots, \mathbf{k}_n), \ldots, \mathrm{Su}(\mathbf{k}_n, \mathbf{k}_1, \ldots, \mathbf{k}_n), \mathbf{z})$. We must show that $\vdash_{\mathrm{PA}} \ulcorner S_i(\mathbf{z}) \urcorner = \mathrm{Su}(\mathbf{k}_i, \mathbf{k}_1, \ldots, \mathbf{k}_n)$. But since $S_i(\mathbf{z})$ is the result of respectively substituting the numerals $\mathbf{k}_1, \ldots, \mathbf{k}_n$ for all free occurrences of $x_1, \ldots, x_n$ in the formula $P_i(\mathrm{Su}(x_1, x_1, \ldots, x_n), \ldots, \mathrm{Su}(x_n, x_1, \ldots, x_n), \mathbf{z})$, which is the formula with Gödel number k_i, we have that the Gödel number of $S_i(\mathbf{z}) = \mathrm{su}(k_i, k_1, \ldots, k_n)$ and, therefore, that $\vdash_{\mathrm{PA}} \ulcorner S_i(\mathbf{z}) \urcorner = \mathrm{Su}(\mathbf{k}_i, \mathbf{k}_1, \ldots, \mathbf{k}_n)$. ⊣

Corollary
For every formula $P(x)$ of PA containing no variable but x free, there exists a sentence S such that $\vdash_{\mathrm{PA}} S \leftrightarrow P(\ulcorner S \urcorner)$.

Proof. This is just the special case of the lemma in which $n = 1$ and $m = 0$. ⊣

If $\vdash_{\mathrm{PA}} S \leftrightarrow P(\ulcorner S \urcorner)$, then S is called a *fixed point* of the predicate $P(x)$. According to the corollary, every predicate has a fixed point.

Theorem 3
If $\vdash_G A$, then for every realization ϕ, $\vdash_{\mathrm{PA}} A^\phi$.

Proof. In view of Theorem 1, it is sufficient to show that for every modal sentence A and every realization ϕ, $\vdash_{\mathrm{PA}} (\Box(\Box A \to A) \to \Box A)^\phi$. Let A be a modal sentence and ϕ a realization. Let $P(x)$ be the formula $(\mathrm{Bew}(x) \to A^\phi)$. By the corollary to the generalized diagonal lemma, there is a sentence S such that $\vdash_{\mathrm{PA}} S \leftrightarrow (\mathrm{Bew}[S]$

$\rightarrow A^{\phi}$). ($\mathrm{Bew}[S] = \mathrm{Bew}(\ulcorner S \urcorner)$.) It follows that $\vdash_{\mathrm{PA}} \mathrm{Bew}[(S \leftrightarrow (\mathrm{Bew}[S] \rightarrow A^{\phi}))]$. Let ψ be a realization such that $\psi(p) = A^{\phi}$ and $\psi(q) = S$. Then $\vdash_{\mathrm{PA}} (\Box(q \leftrightarrow (\Box q \rightarrow p)))^{\psi}$, and $(\Box(\Box p \rightarrow p) \rightarrow \Box p)^{\psi} = \mathrm{Bew}[(\mathrm{Bew}[A^{\phi}] \rightarrow A^{\phi})] \rightarrow \mathrm{Bew}[A^{\phi}] = (\Box(\Box A \rightarrow A) \rightarrow \Box A)^{\phi}$. But by Theorems 1 and 2, $\vdash_{\mathrm{PA}} (\Box(q \leftrightarrow (\Box q \rightarrow p)) \rightarrow (\Box(\Box p \rightarrow p) \rightarrow \Box p))^{\psi}$, whence $\vdash_{\mathrm{PA}} (\Box(\Box A \rightarrow A) \rightarrow \Box A)^{\phi}$. ⊣

One of Solovay's two completeness theorems is that the converse of Theorem 3 is true and, consequently, that a modal sentence A is a theorem of G *if* and only if for every realization ϕ, A^{ϕ} is a theorem of PA.

Theorem 4 gives an interesting alternative characterization of G.

Theorem 4
G is the smallest normal system L of modal logic extending $K4$ the set of whose theorems is closed under the rule of inference: From $\vdash_L \Box A \rightarrow A$, infer $\vdash_L A$.

Proof. Theorem 9 of Chapter 1 showed that G extends $K4$. And if $\vdash_G \Box A \rightarrow A$, then by necessitation, $\vdash_G \Box(\Box A \rightarrow A)$; since $\vdash_G \Box(\Box A \rightarrow A) \rightarrow \Box A$, by modus ponens, $\vdash_G \Box A$, and by modus ponens again, $\vdash_G A$. Suppose that L is a normal system of modal logic extending $K4$, and that for every sentence A, if $\vdash_L \Box A \rightarrow A$, then $\vdash_L A$. We must show that every theorem of G is a theorem of L. To show this it suffices to show that for every sentence A, $\Box(\Box A \rightarrow A) \rightarrow \Box A$ is a theorem of L. Now

(1) $\Box(\Box(\Box A \rightarrow A) \rightarrow \Box A) \rightarrow (\Box\Box(\Box A \rightarrow A) \rightarrow \Box\Box A)$,

(2) $\Box(\Box A \rightarrow A) \rightarrow \Box\Box(\Box A \rightarrow A)$, and

(3) $\Box(\Box A \rightarrow A) \rightarrow (\Box\Box A \rightarrow \Box A)$

are theorems of L, since (1) and (3) are distribution axioms, and (2) is a sentence of the form $\Box B \rightarrow \Box\Box B$.

Then

(4) $\Box(\Box(\Box A \to A) \to \Box A) \to (\Box(\Box A \to A) \to \Box A)$

is also a theorem of L, for it is implied by (2), (1), and (3) in the propositional calculus. But (4) is itself of the form $\Box B \to B$: Take $B = \Box(\Box A \to A) \to \Box A$. $\Box(\Box A \to A) \to \Box A$ is therefore a theorem of L. ⊣

It is not difficult to show that, unlike $K4$, K is closed under the rule of inference: From $\vdash \Box A \to A$, infer $\vdash A$. (Cf. Chapter 8, Exercise 3.) Thus, adjoining the modal-logical "schema" version of Löb's theorem to K yields a strictly stronger system than does adjoining the "rule-of-inference" version. (Adding either to $K4$ yields the same system as adding the other, namely, G.)

Let us now look at some elementary examples of the ways in which a study of G can give us information about provability in arithmetic.

Terminology. Suppose that S and S' are sentences of the language of arithmetic. Then the *arithmetization* of the assertion that

S is provable (in arithmetic) is the sentence Bew[S]
S is consistent (with arithmetic) is the sentence $-$Bew[$-S$]
S is unprovable is the sentence $-$Bew[S]
S is disprovable (refutable) is the sentence Bew[$-S$]
S is undecidable is the sentence ($-$Bew[S] & $-$Bew[$-S$])
S is equivalent to S' is the sentence Bew[$(S \leftrightarrow S')$]
S implies S' (S' is deducible from S) is the sentence Bew[$(S \to S')$]
arithmetic is consistent is the sentence $-$Bew[$\bot$]
arithmetic is inconsistent is the sentence Bew[$\bot$].

The arithmetization of the assertion that if . . . then ____ is the conditional whose antecedent and consequent are the arithmetizations of the assertion that . . . and the assertion that ____ (and similarly for other propositional connectives). An assertion is said to be provable in PA when its arithmetization is. We shall often say 'it is provable that . . .', meaning 'the assertion that . . . is provable' and shall allow ourselves a certain amount of stylistic variation in the choice of expressions with which we refer to assertions, for example, we may use 'the consistency of arithmetic' to refer to the assertion that arithmetic is consistent or we may "anaphorically" use 'it' in place of '*S*', etc.

The Second Incompleteness Theorem of Gödel (for PA) is the assertion that if arithmetic is consistent, then the consistency of arithmetic is not provable in arithmetic. An easy argument, which uses the fact that $\Box(\Box\bot \to \bot) \to \Box\bot$ is a theorem of G, shows that the second incompleteness theorem, which is of course mathematically demonstrable, is in fact *provable in PA*: Since $\vdash_G \Box(\Box\bot \to \bot) \to \Box\bot$, $\vdash_G -\Box\bot \to -\Box-\Box\bot$, and then by Theorem 3, $\vdash_{PA} (-\Box\bot \to -\Box-\Box\bot)^\phi$, that is, $\vdash_{PA} -\text{Bew}[\bot] \to -\text{Bew}[-\text{Bew}[\bot]]$. But this theorem of PA is just the arithmetization of the assertion that if arithmetic is consistent, then the consistency of arithmetic is not provable in arithmetic.

Moreover, $\vdash_G \Box\bot \to \Box\Box\bot$, $\vdash_G -\Box\Box\bot \to -\Box\bot$, and so $\vdash_G -\Box\Box\bot \to (-\Box-\Box\bot \;\&\; -\Box--\Box\bot)$. Therefore the following assertion is provable in PA: If the inconsistency of arithmetic is not provable, then the consistency of arithmetic is undecidable.

A theory *T* whose language is that of PA is said to be *ω-consistent* if there is no formula $A(x)$ such that $\vdash_T \exists x A(x)$ and for every number *n*, $\vdash_T -A(\mathbf{n})$. A sentence *S* in the language of a theory *T* is said to be *undecidable in T* if neither $\vdash_T S$ nor $\vdash_T -S$. And *T* is *incomplete* if there is at least one sentence that is undecidable in *T*. The First Incompleteness Theorem of Gödel is the assertion that if arithmetic is ω-consistent, then arithmetic is incomplete.

A formula $A(x)$ is *decidable* if for every n, either $\vdash_{\mathrm{PA}} A(\mathbf{n})$ or $\vdash_{\mathrm{PA}} -A(\mathbf{n})$. Every primitive recursive formula $A(x)$ is decidable. A theory T in the language of PA is said to be *1-inconsistent* if there is no primitive recursive formula $A(x)$ such that $\vdash_T \exists x A(x)$ and for every number n, $\vdash_T - A(\mathbf{n})$.

If PA is 1-consistent and S is not a theorem of PA, then Bew[S] is not a theorem of PA. For if S is not a theorem, then for every m, $\vdash_{\mathrm{PA}} - \mathrm{Pf}(\mathbf{m}, \ulcorner S \urcorner)$; and since $\mathrm{Pf}(x, \ulcorner S \urcorner)$ is primitive recursive, if PA is 1-consistent, then Bew[S], $= \exists x \mathrm{Pf}(x, \ulcorner S \urcorner)$, is not a theorem either. Thus if PA is 1-consistent, then $\perp$ is not a theorem, Bew[$\perp$] is not a theorem, Bew[Bew[$\perp$]] is not a theorem, ...

The foregoing argument that if PA is 1-consistent, then Bew[$\perp$] is not a theorem can be formalized in PA; it is thus provable in PA that if PA is 1-consistent, then the inconsistency of arithmetic is not provable. As (suitable arithmetizations of) the assertions that if PA is ω-consistent, then PA is 1-consistent, that if PA is 1-consistent, then PA is consistent, that if PA is consistent, then the consistency of arithmetic is not provable, and that if the consistency of arithmetic is undecidable, then PA is incomplete can all be proved in PA, the First Incompleteness Theorem of Gödel can also be proved *in PA*.

PA *is* 1-consistent. (Indeed, PA is ω-consistent. Indeed, every theorem of PA is true.) So none of $\perp$, Bew[$\perp$], Bew[Bew[$\perp$]], . . . is a theorem of PA; by Theorem 3 it follows that none of $\perp$, $\Box\perp$, $\Box\Box\perp$, . . . is a theorem of G.

Löb's theorem states that for every sentence S, if $\vdash_{\mathrm{PA}}$ Bew[S] $\rightarrow S$, then $\vdash_{\mathrm{PA}} S$. Since $\vdash_G \Box(\Box p \rightarrow p) \rightarrow \Box p$, by Theorem 3, for every sentence S and every interpretation ϕ such that $\phi(p) = S$, $\vdash_{\mathrm{PA}} (\Box(\Box p \rightarrow p) \rightarrow \Box p)^{\phi}$;

consequently, for every sentence S, the assertion that S is a theorem of PA if S is deducible from the assertion that S is provable in PA is (not only true, as Löb's theorem states, but also) provable in PA.

Can we prove (in PA) that if arithmetic is consistent, then it is 1-consistent? If we let 1-Con be a suitable arithmetization of the assertion that arithmetic is 1-consistent, we are asking whether $\vdash_{PA} -\text{Bew}[\bot] \rightarrow$ 1-Con. Three paragraphs ago we saw that $\vdash_{PA}$ 1-Con $\rightarrow -\text{Bew}[\text{Bew}[\bot]]$. The answer to our question is, therefore, 'No, on pain of 1-inconsistency'. For if we can prove $\vdash_{PA} -\text{Bew}[\bot] \rightarrow$ 1-Con, then $\vdash_{PA} -\text{Bew}[\bot] \rightarrow -\text{Bew}[\text{Bew}[\bot]]$, and so $\vdash_{PA} \text{Bew}[\text{Bew}[\bot]] \rightarrow \text{Bew}[\bot]$, whence by Löb's theorem, $\vdash_{PA} \text{Bew}[\bot]$, and PA is 1-inconsistent.

A similar argument shows that on pain of 1-inconsistency, we cannot prove that $\vdash_{PA} -\text{Bew}[\text{Bew}[\bot]] \rightarrow$ 1-Con. For $\vdash_{PA}$ 1-Con $\rightarrow -\text{Bew}[\text{Bew}[\text{Bew}[\bot]]]$, and thus if $\vdash_{PA} -\text{Bew}[\text{Bew}[\bot]] \rightarrow$ 1-Con, then by Löb's theorem again, $\vdash_{PA} \text{Bew}[\text{Bew}[\bot]]$, and PA is again 1-inconsistent.[2]

If S is a sentence of the language of PA, then the sentence $\text{Bew}[S] \rightarrow S$ is called *the reflection principle for S*. Löb's theorem thus asserts that for all sentences S, S is provable if the reflection principle for S is provable. No sentence consistent with PA implies all reflection principles: If $\vdash_{PA} S \rightarrow (\text{Bew}[R] \rightarrow R)$ for all sentences R, then $\vdash_{PA} S \rightarrow (\text{Bew}[-S] \rightarrow -S)$, whence by the propositional calculus, $\vdash_{PA} \text{Bew}[-S] \rightarrow -S$, and so by Löb's theorem, $\vdash_{PA} -S$, that is, S is not consistent with PA.

$-\text{Bew}[\bot]$ is, of course, equivalent to the reflection principle $\text{Bew}[\bot] \rightarrow \bot$. And $-\text{Bew}[\text{Bew}[\bot]]$ is equivalent to the conjunction of the reflection principles $\text{Bew}[\text{Bew}[\bot]] \rightarrow \text{Bew}[\bot]$ and $\text{Bew}[\bot] \rightarrow \bot$. But there is no one reflection principle that implies $-\text{Bew}[\text{Bew}[\bot]]$. To see this, we first prove that $\vdash_G \Box((\Box p \rightarrow p) \rightarrow -\Box\Box\bot) \rightarrow \Box\Box\bot$.

Let $A = \Box((\Box p \to p) \to -\Box\Box\bot)$. Since $\vdash_G -\Box\Box\bot \to \Diamond(\Diamond\top \,\&\, \Box\Box\bot)$,

$$\vdash_G (A \,\&\, -\Box\Box\bot) \to \Diamond(\Diamond\top \,\&\, \Box\Box\bot \,\&\, -(\Box p \to p)).$$

And since $\vdash_G \Box\Box\bot \to \Box\Box\Box\bot$,

$$\vdash_G (A \,\&\, -\Box\Box\bot) \to \Diamond(\Diamond\top \,\&\, \Box\Box\Box\bot \,\&\, \Box p),$$
$$\vdash_G (A \,\&\, -\Box\Box\bot) \to \Diamond\Diamond(p \,\&\, \Box\Box\bot),$$
$$\vdash_G (A \,\&\, -\Box\Box\bot) \to \Diamond\Diamond((\Box p \to p) \,\&\, \Box\Box\bot),$$
$$\vdash_G (A \,\&\, -\Box\Box\bot) \to \Diamond\Diamond(-\Box\Box\bot \,\&\, \Box\Box\bot), \text{ and}$$
$$\vdash_G (A \,\&\, -\Box\Box\bot) \to \Diamond\Diamond\bot.$$

Since $\vdash_G \Box\Box\top$, $\vdash_G -\Diamond\Diamond\bot$, and so $\vdash_G A \to \Box\Box\bot$.

Thus if $\vdash_{PA} ((\text{Bew}[S] \to S) \to -\text{Bew}[\text{Bew}[\bot]])$, then $\Box((\Box p \to p) \to -\Box\Box\bot)^\phi$ is true for any ϕ with $\phi(p) = S$; it follows that $\Box\Box\bot^\phi$ is true, hence that Bew[$\bot$] is provable, and PA is 1-inconsistent.

According to Theorem 12 of Chapter 1, $\vdash_G \Box(p \leftrightarrow -\Box p) \to \Box(p \leftrightarrow -\Box\bot)$. Theorem 3 shows that it follows that for every sentence S of the language of arithmetic, it is provable that if S is equivalent to the assertion that S is unprovable, then S is equivalent to the assertion that arithmetic is consistent. Since $\vdash_G \Box(p \leftrightarrow -\Box\bot) \to (\Box p \leftrightarrow \Box-\Box\bot)$ (normality) and $\vdash_G \Box-\Box\bot \leftrightarrow \Box\bot$ (Theorem 11 of Chapter 1), we have $\vdash_G \Box(p \leftrightarrow -\Box p) \to (\Box p \leftrightarrow \Box\bot)$. Thus, for every S, it is provable that if S is equivalent to the assertion that S is unprovable, then S is provable if and only if arithmetic is inconsistent.

By normality, $\vdash_G \Box(p \leftrightarrow \Box p) \to \Box(\Box p \to p)$ and $\vdash_G \Box p \to \Box(p \leftrightarrow \top)$. Since $\vdash_G \Box(\Box p \to p) \to \Box p$, $\vdash_G \Box(p \leftrightarrow \Box p) \to \Box(p \leftrightarrow \top)$. For every sentence S, therefore, it is provable that if S is equivalent to the assertion that S is provable, then S is equivalent to anything that is provable. And since $\vdash_G \Box(p \leftrightarrow \Box p) \to \Box p$, if S is equivalent to the assertion that S is provable, then S is provable.

In like manner, we can prove that $\vdash_G \Box(p \leftrightarrow \Box-p)$

$\rightarrow \Box(p \leftrightarrow \Box\bot)$, and hence that it is provable that if S is equivalent to the assertion that S is disprovable, S is equivalent to the assertion that arithmetic is inconsistent. And since $\vdash_G \Box(p \leftrightarrow -\Box-p) \rightarrow \Box(p \leftrightarrow \bot)$, it is provable that if S is equivalent to the assertion that S is consistent with arithmetic, then S is equivalent to anything that is disprovable.

A conjecture arises. Every occurrence of p in each of $-\Box p$, $\Box p$, $\Box -p$, and $-\Box -p$ is in the scope of some occurrence of $\Box$. Let us call a sentence *modalized in p* if every occurrence of p in that sentence lies in the scope of some occurrence of $\Box$.

Is it the case that for every sentence A containing just p and modalized in p (such as $-\Box p$, $\Box p$, $\Box -p$, and $-\Box -p$), there is a letterless sentence H such that $\vdash_G \Box(p \leftrightarrow A) \rightarrow \Box(p \leftrightarrow H)$? [For $A = -\Box p$ ($\Box p$, $\Box -p$, $-\Box -p$), we may take $H = -\Box\bot$ ($\top$, $\Box\bot$, $\bot$), and – a harder case – for $A = \Box(-p \rightarrow \Box\bot) \rightarrow \Box(p \rightarrow \Box\bot)$, we may take $H = \Diamond\Diamond\top \rightarrow \Diamond\Diamond\Diamond\top$. Thus, if S is equivalent to the assertion that the inconsistency of arithmetic is deducible from S if deducible from the negation of S, then S is equivalent to the assertion that if the consistency of arithmetic is consistent with arithmetic, then so is the consistency of the consistency. The assertion to which S turns out to be equivalent is true but unprovable.] In Chapters 9 and 11 we shall show that the answer is *yes* (the Bernardi–Smoryński theorem). Moreover, the generalization of this theorem to modal sentences A containing sentence letters other than p can also be proved: For every sentence A that is modalized in p, there is a sentence H, containing only sentence letters found in A other than p, such that $\vdash_G \Box(p \leftrightarrow A) \rightarrow \Box(p \leftrightarrow H)$ (the fixed-point theorem).

One fact about G and PA might appear to have been overlooked, or at any rate insufficiently attended to, in the foregoing discussion, namely, that every theorem of

PA is *true* (in the standard model). And because every theorem of PA is true, for every sentence S of the language of arithmetic, if Bew[S] is true, then S is a theorem, and S is thus true. Thus for every interpretation ϕ and every modal sentence A, $(\Box A \rightarrow A)^{\phi}$ is true.

And, of course, if A is a theorem of G, then A^{ϕ} is a theorem of PA, and therefore A^{ϕ} is true.

And, of course, if S and $(S \rightarrow S')$ are true, then S' is true.

We now introduce a modal propositional calculus called G^*. The axioms of G^* are all theorems of G and all sentences of the form $\Box A \rightarrow A$, and the sole rule of inference is modus ponens.

Theorem 5

If $\vdash_{G*} A$, then for every interpretation ϕ, A^{ϕ} is true.

The second completeness theorem of Solovay, also proved in Chapter 12, is that the converse of Theorem 5 is true. Thus the theorems of G^* are precisely those modal sentences all of whose translations are true.

In Chapter 7, and again in Chapter 8, we prove that there is a decision procedure for theoremhood in G. (We shall also prove, in Chapter 12, that there is a decision procedure for theoremhood in G^*.) G^* is thus a recursively axiomatized modal propositional calculus.

G^* is not a normal system of modal logic. Although the set of theorems of G^* is closed under modus ponens, it is *not* closed under necessitation: $(\Box\bot \rightarrow \bot)$ is an axiom of G^* and $((\Box\bot \rightarrow \bot) \rightarrow \Diamond\top)$ is a theorem of K and hence of G and G^*. By modus ponens, $\Diamond\top$ is a theorem of G^*. If $\Box\Diamond\top$ is a theorem of G^*, then $(\Box\Diamond\top)^{\phi}$ is true, and therefore the consistency of arithmetic is provable in arithmetic, which is impossible. Thus $\Diamond\top$ is a theorem of G^*, but $\Box\Diamond\top$ is not, and the theorems of G^* are not closed under necessitation.

They are closed under "possibilification," unlike those of G. ($\vdash_G \top$; $\nvdash_G \Diamond\top$.) If $\vdash_{G*} A$, then from $\vdash_{G*} \Box -A \rightarrow -A$ and $\vdash_{G*} A \rightarrow ((\Box -A \rightarrow -A) \rightarrow -\Box -A)$ (a tautology), we have $\vdash_{G*} -\Box -A$, that is, $\vdash_{G*} \Diamond A$. Thus $\top$, $\Diamond\top$, $\Diamond\Diamond\top$, etc., are all theorems of G^*.

If a correct-modal-principle-when-□-means-'provable' is a modal sentence of which all translations are true, then all theorems of G^* are indeed correct-modal-principles-when-□-means-'provable'. Conversely, too, as we shall see.

There is an interesting "self-strengthening" of Löb's theorem whose proof is our last application of G in this chapter.

Theorem 6
If $\vdash_{\mathbf{PA}}$ Bew[S] & Bew[S'] $\rightarrow S$, then $\vdash_{\mathbf{PA}}$ Bew[S'] $\rightarrow S$.

Proof. We shall show that $\vdash_G \Box(\Box p \,\&\, \Box p' \rightarrow p) \rightarrow \Box(\Box p' \rightarrow p)$. $\vdash_G \Box p' \rightarrow \Box\Box p'$ and $\vdash_G \Box(\Box p' \rightarrow p) \rightarrow (\Box\Box p' \rightarrow \Box p)$, whence by the propositional calculus, $\vdash_G (\Box p \,\&\, \Box p' \rightarrow p) \rightarrow (\Box(\Box p' \rightarrow p) \rightarrow (\Box p' \rightarrow p))$. By normality,

$$\vdash_G \Box(\Box p \,\&\, \Box p' \rightarrow p) \rightarrow \Box(\Box(\Box p' \rightarrow p) \rightarrow (\Box p' \rightarrow p)).$$

But since $\vdash_G \Box(\Box A \rightarrow A) \rightarrow \Box A$,

$$\vdash_G \Box(\Box(\Box p' \rightarrow p) \rightarrow (\Box p' \rightarrow p)) \rightarrow \Box(\Box p' \rightarrow p),$$

and so $\vdash_G \Box(\Box p \,\&\, \Box p' \rightarrow p) \rightarrow \Box(\Box p' \rightarrow p)$. ⊣

Thus S is deducible from the hypothesis that S' is provable if (and only if) S is deducible from the conjunction of the hypotheses that S and S' are provable. Moreover, somewhat surprisingly, for each pair of sentences S and S', the foregoing assertion is provable in PA. By contrast, the semantic techniques whose development we shall begin in Chapter 5 and Solovay's completeness theorem, the proof of which makes crucial use of those techniques, will show us that since $\Box(\Box p \vee \Box -p) \rightarrow (\Box p \vee \Box -p)$ is not a theorem of G (even though it is a theorem of G^*), there exists a sentence S for which the following assertion, although true and unsurprising, is not provable: If it is provable that S is decidable, then S is decidable.

4

Some applications of G

Let us call a sentence of arithmetic *deictic* if it is a member of the smallest class that contains $\perp$ and contains $(S \rightarrow S')$ and Bew[S] whenever it contains S and S'. The class of deictic sentences is a quite natural class to study in proof theory, for it contains (arithmetizations of) a large number of assertions that involve the concepts of provability and consistency. Among the deictic sentences are the arithmetizations of the assertion that arithmetic is consistent, that the consistency of arithmetic is not provable, that if arithmetic is consistent, then it is consistent that it is consistent, etc. The arithmetization of the second incompleteness theorem is also a deictic sentence, of course.

In the previous chapter we saw what the truth-values of some of the simplest deictic sentences are. We shall now see how to calculate the truth-value of any given deictic sentence.[1]

Let us note that for every deictic sentence S there is exactly one letterless sentence A such that $S = A^{\phi}$, and that A^{ϕ} is deictic if A is letterless. (Recall that the denotation of 'A^{ϕ}' depends only on that of 'A' and not on that of 'ϕ' in case the denotation of 'A' is letterless.)

Here are some definitions:

$$\Box^0 A = A; \Box^{n+1} A = \Box\Box^n A.$$

A *lie* is a sentence $\Box^n\bot$, for some $n \geqslant 0$.

A modal sentence is in *normal form* if it is a (possibly empty) conjunction, of which each conjunct is either a lie or a sentence of the form $-\Box^{i+1}\bot \vee \Box^i\bot$. (N.B. Since the empty conjunction is equivalent to $\top$ in the propositional calculus, we shall identify it with $\top$.)

Theorem 1

For every letterless sentence A there is a sentence B in normal form such that $\vdash_G A \leftrightarrow B$.

Proof. Induction on the complexity of letterless sentences. $\bot$ is already in normal form. If $\vdash_G A \leftrightarrow C$, $\vdash_G A' \leftrightarrow C'$, and C and C' are in normal form, then $\vdash_G (A \to A') \leftrightarrow (C \to C')$. To find a B in normal form such that $\vdash_G (A \to A') \leftrightarrow B$, we

(1) Rewrite $(C \to C')$ as a propositional-calculus-equivalent conjunction of disjunctions of lies and negations of lies; in each of the conjoined disjunctions, let all negations come first.

Observe that $\vdash_G \bot \to \Box^i\bot$ for every $i \geqslant 0$ and therefore by normality,

(a) $\vdash_G \Box^m\bot \to \Box^n\bot$ if $m \leqslant n$, and therefore

(b) $\vdash_G (\Box^m\bot \vee \Box^n\bot) \leftrightarrow \Box^n\bot$ if $m \leqslant n$, and

(c) $\vdash_G (-\Box^m\bot \vee -\Box^n\bot) \leftrightarrow -\Box^m\bot$ if $m \leqslant n$.

(2) Using (b), reduce the maximum number of lies in each disjunction to one.

(3) Using (c), reduce the maximum number of negations of lies in each disjunction to one.

(4) Replace each conjunct of the form $-\Box^i\bot$ by $(-\Box^i\bot \vee \Box^0\bot)$.

(5) Using (a), delete all conjuncts of the form $(-\Box^m\bot \vee \Box^n\bot)$, where $m \leqslant n$. If all conjuncts are thereby deleted, we have obtained the empty conjunction and are done. But if not, our B can be obtained by again using (a) to replace each con-

junct of the form $(-\Box^m\bot \vee \Box^n\bot)$, where $m > n$, by $(-\Box^m\bot \vee \Box^{m-1}\bot) \mathbin{\&} (-\Box^{m-1}\bot \vee \Box^{m-2}\bot) \mathbin{\&} \cdots \mathbin{\&} (-\Box^{n+1}\bot \vee \Box^n\bot)$. The result is then in normal form.

Suppose now that $\vdash_G A \leftrightarrow C$, with C in normal form. We want a B in normal form such that $\vdash_G \Box A \leftrightarrow B$. If C is empty, then $\vdash_G A$ and, therefore, $\vdash_G \Box A$; in this case we take B to be the empty conjunction. Otherwise C is a nonempty conjunction $D_1 \mathbin{\&} \cdots \mathbin{\&} D_k$. Since $\vdash_G \Box(D_1 \mathbin{\&} \cdots \mathbin{\&} D_k) \leftrightarrow (\Box D_1 \mathbin{\&} \cdots \mathbin{\&} \Box D_k)$ and $\vdash_G \Box(-\Box^{i+1}\bot \vee \Box^i\bot) \leftrightarrow \Box^{i+1}\bot$, we have that, for some $i_1, \ldots, i_k$, $\vdash_G \Box A \leftrightarrow (\Box^{i_1}\bot \mathbin{\&} \cdots \mathbin{\&} \Box^{i_k}\bot)$. $(\Box^{i_1}\bot \mathbin{\&} \cdots \mathbin{\&} \Box^{i_k}\bot)$ is in normal form. ⊣

$\bot^\phi$, $= \bot$, is false; if C^ϕ is false, C^ϕ is not provable, and then Bew$[C^\phi]$, $= (\Box C)^\phi$, is false. It follows that if D is a lie, D^ϕ is false.

Thus if B is in normal form, B^ϕ is true if and only if no conjunct of B is a lie; for if no conjunct is a lie, every conjunct is a disjunction of which one disjunct is the negation of a lie.

We can therefore effectively decide the truth-value of any given deictic sentence S: Given S, find the letterless A such that $A^\phi = S$. Then, using the procedure described in the proof of the normal form theorem (Theorem 1), find a B in normal form such that $\vdash_G A \leftrightarrow B$. Since B^ϕ has the same truth-value as S, S is true if and only if no conjunct of B is a lie.

If B is nonempty and in normal form, then $\nvdash_{PA} B^\phi$: For if B is nonempty, then since $\Box^i\bot \rightarrow (-\Box^{i+1}\bot \vee \Box^i\bot)$ is a tautology, for some i, $\vdash_G B \rightarrow (\Box\Box^i\bot \rightarrow \Box^i\bot)$; thus if $\vdash_{PA} B^\phi$, then $\vdash_{PA} (\Box\Box^i\bot \rightarrow \Box^i\bot)^\phi$, whence by Löb's theorem, $\vdash_{PA} (\Box^i\bot)^\phi$, and PA is 1-inconsistent, which is absurd.

We can therefore also effectively decide whether any given deictic S is provable: Given S, find the letterless A such that $A^\phi = S$, and find a B in normal form such that $\vdash_G A \leftrightarrow B$. If B is empty, then $\vdash_G A$ and $\vdash_{PA} S$. And if $\vdash_{PA} S$, then since $\vdash_{PA} (A \leftrightarrow B)^\phi$, $\vdash_{PA} B^\phi$, and B is empty. S is thus provable if and only if B is empty. (Alternatively, to decide upon the provability of a deictic sentence S, decide whether or not the deictic sentence Bew[S] is true.)

Suppose that *A is letterless, $\vdash_G A \leftrightarrow B$*, and B is in normal form. If $\vdash_{PA} A^\phi$, then $\vdash_{PA} B^\phi$, B is empty, and $\vdash_G A$; conversely, if $\vdash_G A$, then $\vdash_{PA} A^\phi$. Thus *$\vdash_G A$ if and only if for all realizations ϕ, $\vdash_{PA} A^\phi$*. And *$\vdash_{G*} A$ if and only if for all realizations ϕ, A^ϕ is true:* For every i, $\vdash_{G*} \Diamond^i\top$, and so $\vdash_{G*} -\Box^i\bot$. Since $\vdash_G A \leftrightarrow B$, $\vdash_{G*} A \leftrightarrow B$. Thus if A^ϕ is true, B^ϕ is true, no conjunct of B is a lie, every conjunct of B is a disjunction of which one disjunct is the negation of a lie, $\vdash_{G*} B$, and $\vdash_{G*} A$; conversely, if $\vdash_{G*} A$, then A^ϕ is true.

The two italicized statements are the special "letterless" cases of Solovay's completeness theorems. The proofs of the full theorems, in whose statements the proviso that A is letterless is absent, are far harder.

The next two theorems, which are consequences of the normal form theorem, will be of interest to us when we come to study Rosser's theorem and Rosser sentences. In Chapter 3 we saw that if $A = -\Box\bot$, then $\vdash_G -\Box\Box\bot \rightarrow (-\Box A \;\&\; -\Box -A)$. Is there a letterless sentence A such that $\vdash_G -\Box\bot \rightarrow (-\Box A \;\&\; -\Box -A)$?

Theorem 2
For no letterless A, $\vdash_G -\Box\bot \rightarrow (-\Box A \;\&\; -\Box -A)$.

Proof. Suppose that A is letterless and $\vdash_G -\Box\bot \rightarrow (-\Box A \;\&\; -\Box -A)$. Then $\vdash_G \Box A \rightarrow \Box\bot$ and $\vdash_G \Box -A \rightarrow \Box\bot$. Consider A, which we may suppose to be in normal form. If A is empty, then $\vdash_G A$, $\vdash_G \Box A$, and $\vdash_G \Box\bot$, which is absurd. If every conjunct D of A is either $-\Box^{i+1}\bot \vee \Box^i\bot$, or $\Box^i\bot$, for some $i \geqslant 1$, then for every conjunct D, $\vdash_G \Box\bot \rightarrow D$, $\vdash_G \Box\Box\bot \rightarrow \Box D$, and therefore $\vdash_G \Box\Box\bot \rightarrow \Box A$, hence $\vdash_G \Box\Box\bot \rightarrow \Box\bot$, which is absurd. But if $\bot$ is a conjunct, then $\vdash_G -A$, $\vdash_G \Box - A$, and $\vdash_G \Box\bot$, which is also absurd. Thus $-\Box\bot \vee \bot$ is a conjunct of A, and therefore $\vdash_G A \rightarrow -\Box\bot$. In like manner, $\vdash_G -A \rightarrow -\Box\bot$, and so $\vdash_G -\Box\bot$, contradiction. ⊣

Theorem 3 tells us that no nonprovable deictic sentence is strictly weaker than $-$Bew[$\bot$].

Theorem 3
Suppose that A is letterless, $\vdash_G -\Box\bot \rightarrow A$, and $\nvdash_G A$. Then $\vdash_G A \leftrightarrow -\Box\bot$.

Proof. We may assume that A is in normal form. Since $\nvdash_G A$, A is a nonempty conjunction of lies and sentences of the form $-\Box^{i+1}\bot \vee \Box^i\bot$. Let D be a conjunct of A. Then $\vdash_G -\Box\bot \rightarrow D$, and so $\vdash_G \Box\bot \vee D$. If $D = \bot$, then $\vdash_G \Box\bot$, which is impossible. If $D = \Box^i\bot$ and $i \geqslant 1$, then $\vdash_G \Box^i\bot$, which is also impossible. Thus D is not a lie. If $D = -\Box^{i+1}\bot \vee \Box^i\bot$ and $i \geqslant 1$, then since $\vdash_G \Box\bot \rightarrow \Box^i\bot$, $\vdash_G \Box^{i+1}\bot \rightarrow \Box^i\bot$, whence $\vdash_G \Box^i\bot$, which is impossible. So $D = -\Box\bot \vee \bot$. But since $\vdash_G D \rightarrow -\Box\bot$, $\vdash_G A \leftrightarrow -\Box\bot$. ⊣

We turn now to the topic of fixed points of predicates.

A *Gödel sentence* is a sentence S such that $\vdash_{PA} S \leftrightarrow -$Bew[$S$]. Since $\Box(p \leftrightarrow -\Box p)^\phi$ is true if and only if $\phi(p) \leftrightarrow -$Bew[$\phi(p)$] is provable, S is a Gödel sentence if and only if for some interpretation ϕ, $\phi(p) = S$ and $\Box(p \leftrightarrow -\Box p)^\phi$ is true. A *Henkin sentence* is a sentence S such that $\vdash_{PA} S \leftrightarrow$ Bew[S]. Thus S is a Henkin sentence if and only if for some ϕ, $\phi(p) = S$ and $\Box(p \leftrightarrow \Box p)^\phi$ is true. Jeroslow has studied sentences S such that $\vdash_{PA} S \leftrightarrow$ Bew[$-S$], that is, sentences S such that for some ϕ, $\phi(p) = S$ and $\Box(p \leftrightarrow \Box - p)^\phi$

is true,[2] and Rogers has discussed sentences S such that $\vdash_{PA} S \leftrightarrow -\text{Bew}[-S]$, that is, sentences S such that for some ϕ, $\phi(p) = S$ and $\Box(p \leftrightarrow -\Box -p)^{\phi}$ is true.[3]

A modal sentence A is called *modalized in p* if each occurrence of p in A lies in the scope of some occurrence of $\Box$. Thus $-\Box p$, $\Box p$, $\Box -p$, and $-\Box -p$ are all modalized in p.

Suppose that A contains no sentence letters other than p. We shall call a sentence S an *A fixed point* if for some interpretation ϕ, $\phi(p) = S$ *and* $\Box(p \leftrightarrow A)^{\phi}$ is true. Gödel sentences are thus $-\Box p$ fixed points; Henkin, $\Box p$; Jeroslow, $\Box -p$; and Rogers, $-\Box -p$. If A is not modalized in p, we cannot in general expect A fixed points to exist: If $A = -p$, then an A fixed point is a sentence S such that $S \leftrightarrow -S$ is provable; thus $-p$ fixed points exist if and only if arithmetic is inconsistent (which it is not). And even when A fixed points do exist for unmodalized A, we cannot in general expect them to be equivalent: If $A = p$, then $\top$ and $\bot$ are inequivalent p fixed points. But if A is modalized in p, then not only do A fixed points always exist (as we shall see shortly) and not only are any two A fixed points equivalent, but (as we shall see in Chapter 9) there exists a *deictic* sentence that is an A fixed point and to which all A fixed points are equivalent (the Bernardi-Smoryński theorem).[4]

We shall call a sentence S a *Gödelian fixed point* if for some sentence A modalized in p and containing no sentence letters but p, S is an A fixed point. Thus Gödel, Henkin, Jeroslow, and Rogers sentences are all Gödelian fixed points.

At the end of this chapter we shall prove a theorem about G that tells us under what conditions Gödelian fixed points are provable. But first we shall prove a theorem that guarantees us that for every A modalized in p and containing no letters but p, an A fixed point exists.

Theorem 4
For every sentence A modalized in p and every interpretation ϕ, there exists a predicate $(A,\phi)(x)$ such that for every sentence S, S is a fixed point of $(A,\phi)(x)$ if and only if for some interpretation ψ, $\psi(p) = S$, $\Box(p \leftrightarrow A)^{\psi}$ is true, and $\psi(r) = \phi(r)$ for all sentence letters $r \neq p$.

Proof. There exist two primitive recursive functions, cond and bw, such that for all m,n, if m and n are the Gödel numbers of F and G, then cond(m,n) and bw(m) are the Gödel numbers of $(F \rightarrow G)$ and Bew$[F]$. Let Cond(x_1,x_2) and Bw(x_1) be primitive recursive terms for cond and bw. Then for all F, G, $\vdash_{\text{PA}}$ Cond$(\ulcorner F\urcorner,\ulcorner G\urcorner)$ $= \ulcorner(F \rightarrow G)\urcorner$ and $\vdash_{\text{PA}}$ Bw$(\ulcorner F\urcorner) = \ulcorner$Bew$[F]\urcorner$.

We now assign to each modal sentence C (whether modalized in p or not) and each interpretation ϕ a *term* $(D/\phi)(x)$, as follows:

> $(\bot/\phi)(x)$ is $\ulcorner\bot\urcorner$;
> $(p/\phi)(x)$ is the variable x;
> $(r/\phi)(x)$ is $\ulcorner\phi(r)\urcorner$ if r is a sentence letter other than p;
> $((B \rightarrow C)/\phi)(x)$ is Cond$((B/\phi)(x), (C/\phi)(x))$; and
> $(\Box B/\phi)(x)$ is Bw$((B/\phi)(x))$.

A mechanical induction on the complexity of modal sentences establishes that if $\phi(p) = S$, then for every modal sentence D, $\vdash_{\text{PA}} (D/\phi)(\ulcorner S\urcorner) = \ulcorner D^{\phi}\urcorner$. We do the $\Box B$ case: By the induction hypothesis, $\vdash_{\text{PA}} (B/\phi)(\ulcorner S\urcorner)$ $= \ulcorner B^{\phi}\urcorner$; therefore $\vdash_{\text{PA}} (\Box B/\phi)(\ulcorner S\urcorner) =$ Bw$((B/\phi)(\ulcorner S\urcorner))$ $=$ Bw$(\ulcorner B^{\phi}\urcorner) = \ulcorner$Bew$[B^{\phi}]\urcorner = \ulcorner(\Box B)^{\phi}\urcorner$. An equally mechanical induction shows that if $\psi(r) = \phi(r)$ for all sentence letters $r \neq p$, then $(D/\psi)(x)$ is the same term as $(D/\phi)(x)$.

We now define $(A,\phi)(x)$ for A modalized in p:

> Let $(\Box D,\phi)(x)$ be Bew$((D/\phi)(x))$ (for all modal sentences D);
> let $(\bot,\phi)(x)$ be $\bot$;

let $(r,\phi)(x)$ be $\phi(r)$ if r is a sentence letter other than p; and

let $((B \to C),\phi)(x)$ be $((B,\phi)(x) \to (C,\phi)(x))$.

If A is modalized in p, then A is a truth-functional combination of sentences of the form $\Box D$ and sentence letters other than p, and therefore $(A,\phi)(x)$ is defined. If $\phi(p) = S$, then, since $\vdash_{PA} (D/\phi)(\ulcorner S \urcorner) = \ulcorner D^{\phi} \urcorner$, $\vdash_{PA} (\Box D,\phi)(\ulcorner S \urcorner) \leftrightarrow \mathrm{Bew}(\ulcorner D^{\phi} \urcorner) \leftrightarrow (\Box D)^{\phi}$, and then it follows by a trivial truth-functional induction that for every sentence A modalized in p, $\vdash_{PA} (A,\phi)(\ulcorner S \urcorner) \leftrightarrow A^{\phi}$. And, as in the case of terms $(D/\phi)(x)$, if $\psi(r) = \phi(r)$ for all sentence letters $r \neq p$, then $(A,\psi)(x)$ is the same predicate as $(A,\phi)(x)$.

Suppose now that S is a fixed point of $(A,\phi)(x)$. Let $\psi(p) = S$, $\psi(r) = \phi(r)$ for all sentence letters $r \neq p$. We must show that $\Box(p \leftrightarrow A)^{\psi}$ is true. But $(A,\psi)(x)$ is the same predicate as $(A,\phi)(x)$, and therefore $\vdash_{PA} S \leftrightarrow (A,\psi)(\ulcorner S \urcorner)$. And we have seen that $\vdash_{PA} (A,\psi)(\ulcorner S \urcorner) \leftrightarrow A^{\psi}$. Thus $S \leftrightarrow A^{\psi}$ is provable and so $\Box(p \leftrightarrow A)^{\psi}$ is true. Conversely, if $\psi(p) = S$, $\Box(p \leftrightarrow A)^{\psi}$ is true, and $\psi(r) = \phi(r)$ for all sentence letters $r \neq p$, then $\vdash_{PA} S \leftrightarrow A^{\psi}$, whence, since $\vdash_{PA} (A,\psi)(\ulcorner S \urcorner) \leftrightarrow A^{\psi}$, $\vdash_{PA} S \leftrightarrow (A,\psi)(\ulcorner S \urcorner)$; but since $(A,\phi)(x)$ is the same predicate as $(A,\psi)(x)$, the theorem is proved. $\dashv$

If we let $\mathrm{Neg}(x)$ be the term $\mathrm{Cond}(x,\ulcorner \bot \urcorner)$, then it is easy to see that, for example, $(\Box p,\phi)(x)$ is $\mathrm{Bew}(x)$, $(-\Box - p, \phi)(x)$ is $-\mathrm{Bew}(\mathrm{Neg}(x))$, $(\Box - - p \to - \Box - \Box p, \phi)(x)$ is $(\mathrm{Bew}(\mathrm{Neg}(\mathrm{Neg}(x))) \to -\mathrm{Bew}(\mathrm{Neg}(\mathrm{Bw}(x))))$, etc.

Corollary

If A is modalized in p and contains no letters other than p, then there is at least one A fixed point.

Proof. Let ϕ be some interpretation. By the corollary to the generalized diagonal lemma, a fixed point S of $(A,\phi)(x)$ exists. By Theorem 4, for some interpretation

ψ, $\psi(p) = S$ and $\Box(p \leftrightarrow A)^{\psi}$ is true, that is, S is an A fixed point. $\dashv$

The corollary to Theorem 5 will tell us what the conditions are under which a Gödelian fixed point is provable. (We shall use '$A(p,\mathsf{q})$' and '$A(p)$' to denote sentences containing no letters except p,q, and sentences containing no letters but p. 'q' abbreviates '$q_1, \ldots, q_s$'.)

Theorem 5
Suppose that $A(p,\mathsf{q})$ is modalized in p. Then
$\vdash_G \Box(p \leftrightarrow A(p,\mathsf{q})) \rightarrow (\Box p \leftrightarrow \Box A(\top,\mathsf{q}))$.

Proof. Since $A(p,\mathsf{q})$ is modalized in p, there exist sentences $A_1(p,\mathsf{q}), \ldots, A_n(p,\mathsf{q})$ such that $A(p,\mathsf{q})$ is a truth-functional combination of $\Box A_1(p,\mathsf{q}), \ldots, \Box A_n(p,\mathsf{q})$, and q. By Theorems 5 and 9 of Chapter 1, for each i, $1 \leqslant i \leqslant n$, $\vdash_G \Box(p \leftrightarrow \top) \rightarrow \Box(A_i(p,\mathsf{q}) \leftrightarrow A_i(\top,\mathsf{q}))$, and by normality, $\vdash_G \Box p \rightarrow \Box(p \leftrightarrow \top)$, and $\vdash_G \Box p \rightarrow (\Box A_i(p,\mathsf{q}) \leftrightarrow \Box A_i(\top,\mathsf{q}))$. By the propositional calculus, then,

(1) $\vdash_G \Box p \rightarrow (A(p,\mathsf{q}) \leftrightarrow A(\top,\mathsf{q}))$.

From (1) by normality we have $\vdash_G \Box\Box p \rightarrow (\Box A(p,\mathsf{q}) \rightarrow \Box A(\top,\mathsf{q}))$. By normality, we also have $\vdash_G \Box(p \leftrightarrow A(p,\mathsf{q})) \rightarrow (\Box p \rightarrow \Box A(p,\mathsf{q}))$. Since $\vdash_G \Box p \rightarrow \Box\Box p$, by the propositional calculus, we have

(2) $\vdash_G \Box(p \leftrightarrow A(p,\mathsf{q})) \rightarrow (\Box p \rightarrow \Box A(\top,\mathsf{q}))$.

On the other hand, by (1) we have $\vdash_G A(\top,\mathsf{q}) \rightarrow (\Box p \rightarrow A(p,\mathsf{q}))$, whence by normality, $\vdash_G \Box A(\top,\mathsf{q}) \rightarrow \Box(\Box p \rightarrow A(p,\mathsf{q}))$. Again by normality,

$$\vdash_G \Box(p \leftrightarrow A(p,\mathsf{q})) \rightarrow (\Box(\Box p \rightarrow A(p,\mathsf{q})) \rightarrow \Box(\Box p \rightarrow p)).$$

Thus $\vdash_G \Box(p \leftrightarrow A(p,\mathsf{q})) \rightarrow (\Box A(\top,\mathsf{q}) \rightarrow \Box(\Box p \rightarrow p))$. But $\vdash_G \Box(\Box p \rightarrow p) \rightarrow \Box p$, whence $\vdash_G \Box(p \leftrightarrow A(p,\mathsf{q})) \rightarrow (\Box A(\top,\mathsf{q}) \rightarrow \Box p)$, which, together with (2), yields the theorem. $\dashv$

Corollary
Suppose that $A(p)$ is modalized in p. Then $\vdash_G \Box(p \leftrightarrow A(p)) \to (\Box p \leftrightarrow \Box A(\top))$.

Suppose now that $A(p)$ is modalized in p and S is an $A(p)$ fixed point. Then for some interpretation ϕ, $\phi(p) = S$ and $\Box(p \leftrightarrow A(p))^\phi$ is true. By the corollary to Theorem 5, $\vdash_{\mathbf{PA}} \Box(p \leftrightarrow A(p))^\phi \to ((\Box p)^\phi \leftrightarrow (\Box A(\top))^\phi)$, and so $\Box(p \leftrightarrow A(p))^\phi \to (\mathrm{Bew}[S] \leftrightarrow \mathrm{Bew}[A(\top)^\phi])$ is true, and therefore S is provable if and only if $A(\top)^\phi$, which is a deictic sentence, is provable. Therefore, if $A(p)$ is modalized in p, each $A(p)$ fixed point is provable if and only if all $A(p)$ fixed points are provable.

It follows that a Gödel sentence is provable if and only if $-\mathrm{Bew}[\top]$ is provable. (Here, 'A' may mean 'every' or 'some'; it does not matter which.) Since $\mathrm{Bew}[\top]$ is provable, a Gödel sentence is provable if and only if arithmetic is inconsistent. Similarly, a Henkin sentence is provable if and only if $\mathrm{Bew}[\top]$ is provable; thus all Henkin sentences are provable (as Löb showed). A Jeroslow sentence is provable if and only if $\mathrm{Bew}[-\top]$ is provable, hence if and only if the inconsistency of arithmetic is provable. And a Rogers sentence is provable if and only if $-\mathrm{Bew}[-\top]$ is provable, if and only if the consistency of arithmetic is provable, if and only if arithmetic is inconsistent.

A few pages ago we gave an effective procedure for deciding whether or not any given deictic sentence is provable. We have now established that there is an effective procedure for deciding whether or not any given Gödelian fixed point S is provable: Find an $A(p)$, modalized in p, such that S is an $A(p)$ fixed point (this can, of course, be done effectively) and decide whether or not the deictic sentence $A(\top)^\phi$ is provable (perhaps by putting $A(\top)$ in normal form and seeing whether the result is empty); S is provable iff $A(\top)^\phi$ is.

Having seen what the corollary to Theorem 5 tells us about the provability in arithmetic of Gödelian fixed points, we may perhaps remark that Theorem 5 tells us that every fixed point of $(A(p,\mathsf{q}),\phi)(x)$ is provable if and only if $A(\top,\mathsf{q})^{\phi}$ is provable. Notice the uniformity: There is an effective operation X on *sentences* $A(p,\mathsf{q})$ modalized in p such that for every $A(p,\mathsf{q})$, a fixed point of $(A(p,\mathsf{q}),\phi)(x)$ is provable if and only if $\mathrm{X}(A(p,\mathsf{q}))^{\phi}$ is provable, for all realizations ϕ.

5

Semantics for *G* and other modal logics

The semantical treatment of modal logic that we now present is due to Kripke and was inspired by a well-known fantasy of Leibniz, according to which we inhabit a place called *the actual world,* which is one of a number of *possible worlds.* (The world we inhabit was thought by God to have certain excellences with the enumeration of which it is not necessary to trouble ourselves, and because of these our world was selected by God to be the possible world that He would make actual.) Each of our statements is true or false in – we shall say *at* – various possible worlds. A statement is true at a world if it correctly describes that world, and false if it does not. We sometimes call a particular statement true or false, *tout court,* but when we do, we are to be understood as speaking about the actual world, and saying that the statement is true or false *at it.* Some of the statements we make are true at *all* possible worlds, including of course the actual world; these are the so-called *necessary* statements. A statement to the effect that another is necessary will thus be true if the other statement is true at all possible worlds. It follows that if a statement is necessary, then it is true. Some statements are true at at least one possible world; these are the *possible* statements. Since what is true at the actual world is true at at least one possible world, whatever is true is possible. A statement is necessary if and only if its negation is not possible, for the negation of a statement will be true at precisely those worlds at which the statement is false. And if a conditional and its antecedent are both necessary, then the consequent of the conditional is necessary too.

There is a question, raised by Kripke, to which this description of Leibniz's system of possible worlds does not supply the answer. We are said to inhabit the actual world. Are the other possible worlds of whose existence we have been apprised absolutely all of the other worlds that there really are, or are they only those that are possible *relative to* the actual world? The description leaves open whether or not, if we had inhabited some other world than the actual world, there might have been worlds other than those we now acknowledge that were possible *relative to* that other possible world; in brief, our description does not answer the question whether or not exactly the same worlds are possible relative to each possible world that are possible relative to the actual world.

A possible world is called *accessible from* another if it is possible relative to that other. If we do not assume that the worlds accessible from the actual world are precisely the worlds accessible from each world – even though it may appear self-evident that they are – then questions arise about the nature of the accessibility relation. For example, is the relation transitive? If so, then all worlds accessible from worlds that are accessible from the actual world will themselves be accessible from the actual world. It follows that if a statement A is necessary, then A will be true at all worlds x accessible from the actual world; and therefore A will be true at every world y that is accessible from some world x accessible from the actual world (for all such worlds y are accessible from the actual world if accessibility is transitive); and therefore the statement that A is necessary will be true at every world x accessible from the actual world; and therefore the statement that A is necessary will itself be necessary. Thus, on the assumption that the accessibility relation is transitive, if a statement A is necessary, then the statement that A is necessary will also be necessary. In like manner other determinations of the character of the accessibility relation can guarantee the correctness of other modal principles. (The system of semantics for G that we shall give in this chapter will differ from Leibniz's system in that no world will ever be accessible from itself!)

Set-theoretical analogues of these metaphysical notions were defined by Kripke in providing what has become the standard sort of model-theoretic semantics for the most common systems of modal propositional logic.[1] Here are the definitions of some standard concepts of modal-logical semantics.

A *frame* is an ordered pair $\langle W, R\rangle$ consisting of a non-empty set W and a binary relation R on W, that is, a set of ordered pairs of members of W. P is an *evaluator* on W if P is a function that assigns a truth-value, $\top$ or $\bot$, to each ordered pair consisting of a member of W and a sentence letter. If $\langle W, R\rangle$ is a frame and P is an evaluator on W, then the ordered triple $\langle W, R, P\rangle$ is said to be a *model based on* $\langle W, R\rangle$. The members of W are called the *possible worlds* (or sometimes just the *worlds*) of both the frame $\langle W, R\rangle$ and the model $\langle W, R, P\rangle$, and W and R are called the *domain* and the *accessibility relation* of both $\langle W, R\rangle$ and $\langle W, R, P\rangle$. A model or a frame will sometimes be said to have some property of binary relations, for example, transitivity, if its accessibility relation has that property.

Modal sentences will be said to be true or false *at* worlds *in* models. (A world at which a sentence is true or false in a model is always a member of the domain of the model.) The function of an evaluator is to determine in the obvious way which "atomic sentences," that is, which sentence letters, are to be true at which worlds in a model. The truth-value at a world of a truth-functional compound is to be determined in the usual manner by the truth-values at the same world of its components. A necessitation of a sentence is to be true at a world just in case the sentence is true at all worlds accessible from the world. Thus, writing '$\langle W,R,P\rangle \vDash_w A$' to mean '$A$ is true at w in $\langle W,R,P\rangle$' and '$\langle W,R,P\rangle \nvDash_w A$' to mean '$A$ is false at w in $\langle W,R,P\rangle$', we make this definition:

(1) $\langle W,R,P\rangle \nvDash_w \bot$;

(2) $\langle W,R,P\rangle \vDash_w p$ if $P(w,p) = \top$;
$\langle W,R,P\rangle \nvDash_w p$ if $P(w,p) = \bot$
(for all sentence letters p);

(3) $\langle W,R,P\rangle \vDash_w (A \rightarrow B)$ if either $\langle W,R,P\rangle \nvDash_w A$ or $\langle W,R,P\rangle \vDash_w B$;
$\langle W,R,P\rangle \nvDash_w (A \rightarrow B)$ if both $\langle W,R,P\rangle \vDash_w A$ and $\langle W,R,P\rangle \nvDash_w B$;

(4) $\langle W,R,P\rangle \vDash_w \Box A$ if for every x such that wRx, $\langle W,R,P\rangle \vDash_x A$;
$\langle W,R,P\rangle \nvDash_w \Box A$ if for some x such that wRx, $\langle W,R,P\rangle \nvDash_x A$.

(Both '$\langle W,R,P\rangle \vDash_w A$' and '$\langle W,R,P\rangle \nvDash_w A$' are to be undefined if w is not in W.)

Some obvious and familiar consequences of this definition are

$\langle W,R,P\rangle \vDash_w -A$ if $\langle W,R,P\rangle \nvDash_w A$;
$\langle W,R,P\rangle \nvDash_w -A$ if $\langle W,R,P\rangle \vDash_w A$;
$\langle W,R,P\rangle \vDash_w (A \vee B)$ if either $\langle W,R,P\rangle \vDash_w A$ or $\langle W,R,P\rangle \vDash_w B$;
$\langle W,R,P\rangle \nvDash_w (A \vee B)$ if both $\langle W,R,P\rangle \nvDash_w A$ and $\langle W,R,P\rangle \nvDash_w B$;

and similarly for the other propositional connectives.

$\langle W,R,P\rangle \vDash_w \Diamond A$ if for some x such that wRx, $\langle W,R,P\rangle \vDash_x A$; and
$\langle W,R,P\rangle \nvDash_w \Diamond A$ if for every x such that wRx, $\langle W,R,P\rangle \nvDash_x A$.

A sentence is said to be *valid in a model* if it is true at all worlds in the model. And a sentence is said to be *valid in a frame* if it is valid in all models based on the frame. Thus a sentence is valid in all frames iff valid in all models, valid in all transitive frames iff valid in all transitive models, etc.

Every tautology is true at each world in each model; therefore all tautologies are valid in all models. And every distribution axiom is likewise valid in all models because it is true at each world in each model: For if $\langle W, R, P\rangle \models_w \Box(A \rightarrow A')$ and $\langle W, R, P\rangle \models_w \Box A$, then for all x such that wRx, both $\langle W, R, P\rangle \models_x A \rightarrow A'$ and $\langle W, R, P\rangle \models_x A$, and, therefore, for all x such that wRx, $\langle W, R, P\rangle \models_x A'$, that is, $\langle W, R, P\rangle \models_w \Box A'$. If A and $(A \rightarrow B)$ are valid in a model, then so is B. And even though some sentences are true at worlds in models at which their necessitations are false, if a sentence A is *valid in a model,* then so is $\Box A$; for if A is valid in $\langle W, R, P\rangle$, then A is true at every world in $\langle W, R, P\rangle$, therefore A is true at every world x such that wRx, for every world w, that is, $\Box A$ is true at every world w in $\langle W, R, P\rangle$. It follows that every sentence that occurs in every proof in K is valid in all models, and thus that every theorem of K is valid in all models, or, equivalently, valid in all frames. (In the next chapter we shall establish the converse, that every modal sentence that is valid in all models is a theorem of K.)

(It is not true that if a sentence is valid in a model, then every substitution instance of it is valid in the model: q is a substitution instance of p, and p is valid and q invalid in every model $\langle W, R, P\rangle$ in which $P(w,p) = \top$ and $P(w,q) = \bot$ for all worlds w in W. What *is* true is that if A is valid in a *frame* $\langle W, R\rangle$ and B is a substitution instance of A (obtained by replacing p by C, say), then B is valid in $\langle W, R\rangle$: For if B is not valid in $\langle W, R\rangle$, then for some evaluator P on W and some w in W, $\langle W, R, P\rangle \not\models_w B$. Let us define an evaluator Q on W by setting $Q(y,p) = \top$ iff $\langle W,R,P\rangle \vDash_y C$ and $Q(y,q) = P(y,q)$ for $q \neq p$. A completely trivial induction on sentences shows that for every pair of sentences D, E and every y in W, if E is the result of everywhere replacing p by C in D, then $\langle W,R,Q\rangle \vDash_y D$ iff $\langle W,R,P\rangle \vDash_y E$. Since $\langle W,R,P\rangle \nvDash_w B$, $\langle W,R,Q\rangle \nvDash_w A$, and A is not valid in $\langle W,R\rangle$.)

In a moment we shall begin our examination of the conditions under which sentences are valid in frames, and find out what those conditions are in the case of our special axiom $\Box(\Box p \rightarrow p) \rightarrow \Box p$. But we first want to prove a theorem about what the truth-value of a sentence at a world is dependent on. In order to state it, we need to recall the concept of the *ancestral* R_* of a binary relation R.

We define R_* by: xR_*y iff $x = z_1Rz_2 \cdots Rz_n = y$, for some $z_1, \ldots, z_n$.

R is evidently a subrelation of R_*; that is, if xRy, then xR_*y. R_* is evidently transitive, for if xR_*y and yR_*z, then $x = a_1Ra_2 \cdots Ra_m = y = b_1Rb_2R \cdots Rb_n = z$, for some $a_1, \ldots, a_m, b_1, \ldots, b_n$, and so xR_*z. And if R is a subrelation of a transitive relation S, then R_* is a subrelation of S too, for if $z_1R_*z_n$, then $z_1Rz_2R \cdots Rz_n$, and so $z_1Sz_2S \cdots Sz_n$, and therefore z_1Sz_n.

Thus R_* is a transitive relation that includes R and is included in every transitive relation that includes R. It is therefore *the unique* transitive relation that includes R and is included in every transitive relation that includes R. Thus if R is itself transitive, $R_* = R$.

Theorem 1 states that the truth-value of a sentence A at a world w depends only on the truth-values, at w and all other worlds x such that wR_*x, of those sentence letters that are contained in A.

Theorem 1

Suppose that $W' = \{w\} \cup \{x|wR_*x\}$, R' is a relation on W', P' is an evaluator on W', $yR'z$ iff yRz for all y, z in

W', and $P'(y,p) = P(y,p)$ for all y in W' and all sentence letters p contained in A. Then $\langle W,R,P\rangle \models_w A$ iff $\langle W',R',P'\rangle \models_w A$.

Proof. We shall show, by induction on subsentences B of A, that $\langle W,R,P\rangle \models_y B$ iff $\langle W',R',P'\rangle \models_y B$ for every y in W' and every subsentence B of A. The only nontrivial case is that in which $B = \Box C$. Now, when $y \in W'$, if yRz, then yR_*z, wR_*z, $z \in W'$, and by the hypothesis, $yR'z$; and if $yR'z$, then y, $z \in W'$, and so by the hypothesis, yRz. Thus when $y \in W'$, yRz iff $yR'z$. Thus $\langle W,R,P\rangle \models_y \Box C$ iff $\langle W,R,P\rangle \models_z C$ for every z such that yRz, iff $\langle W,R,P\rangle \models_z C$ for every z such that $yR'z$, iff, by the inductive hypothesis, $\langle W',R',P'\rangle \models_z C$ for every z such that $yR'z$, iff $\langle W',R',P'\rangle \models_y \Box C$. ⊣

Theorem 1 has a corollary of which we shall make immediate use.

Corollary
If for every y in W and every p contained in A, $P(y,p) = Q(y,p)$, then $\langle W,R,P\rangle \models_w A$ iff $\langle W,R,Q\rangle \models_w A$.

Proof. Let $W' = \{w\} \cup \{x|wR_*x\}$. Define R' and P' by $yR'z$ iff y, $z \in W'$ and yRz; $P'(y,p) = P(y,p)$ for all y in W'. Then for every y in W' and every sentence letter p contained in A, $P'(y,p) = Q(y,p)$, and by Theorem 1 $\langle W,R,P\rangle \models_w A$ iff $\langle W',R',P'\rangle \models_w A$ iff $\langle W,R,Q\rangle \models_w A$.

We have seen that all theorems of K are valid in all frames. Under what conditions is $\Box p \to p$ valid in a frame $\langle W,R\rangle$? Trivially, $\Box p \to p$ is valid in $\langle W,R\rangle$ if and only if for all evaluators P on W and all w in W, $\langle W,R,P\rangle \models_w \Box p \to p$. But by the corollary to Theorem 1, if Q is an evaluator on W that *everywhere falsifies all letters but p* [that is, if for every y in W and every $q \neq p$, $Q(y,q) = \bot$] and for every y in W, $Q(y,p) = P(y,p)$, then $\langle W,R,P\rangle \models_w \Box p \to p$ iff $\langle W,R,Q\rangle \models_w \Box p \to p$. So

$\Box p \to p$ is valid in $\langle W, R\rangle$ iff for all evaluators P on W that everywhere falsify all letters but p and all w in W, $\langle W, R, P\rangle \vDash_w \Box p \to p$. And $\langle W, R, P\rangle \vDash_w \Box p \to p$ iff, if for all x such that wRx, $P(x,p) = \top$, then $P(w,p) = \top$. There is a "canonical" one-to-one correspondence $\sim$ between subsets S of W and evaluators P on W that everywhere falsify all letters but p, which is defined by: $S \sim P$ iff for every y in W, $y \in S$ iff $P(y,p) = \top$. We can therefore conclude that $\Box p \to p$ is valid in $\langle W, R\rangle$ if and only if *for all subsets S of W and all w in W, if for all x such that wRx, $x \in S$, then $w \in S$.*

In the present instance, we can simplify the right-hand side of the last biconditional, which holds if and only if R is a *reflexive* relation on W. For if R is a reflexive relation on W, $S \subseteq W$, $w \in W$, and for all x such that wRx, $x \in S$, then, by reflexivity wRw, and so $w \in S$. And if $w \in W$, $S = \{x|wRx\}$, and the right-hand side holds, then since for all x such that wRx, $x \in S$, we have that $w \in S$, that is, wRw.

We have thus shown that $\Box p \to p$ is valid in $\langle W, R\rangle$ if and only if R is a reflexive relation on W. What about $\Box p \to \Box\Box p$? As before, we can directly "read off" from $\Box p \to \Box\Box p$ that it is valid in $\langle W, R\rangle$ if and only if

$$\forall S \subseteq W \forall w \in W(\forall x(wRx \to x \in S) \to \forall x(wRx \to \forall y(xRy \to y \in S))).$$

And again we can simplify the right-hand side considerably, which evidently holds in case R is transitive. If, conversely, the right-hand side holds, wRx, and xRy, then wRy too, as we can see by taking $S = \{a|wRa\}$. Thus $\Box p \to \Box\Box p$ is valid in $\langle W, R\rangle$ iff R is a *transitive* relation on W.

Also, $p \to \Box\Diamond p$ is valid in $\langle W, R\rangle$ if and only if

$$\forall S \subseteq W \forall w \in W(w \in S \to \forall x(wRx \to \exists y(xRy \;\&\; y \in S))).$$

Here the right-hand side clearly holds if R is symmetric.

And if the right-hand side holds and wRx, then xRw: Take $S = \{w\}$. So $p \rightarrow \Box\Diamond p$ is valid in $\langle W, R \rangle$ iff R is a *symmetric* relation on W.

And $\Diamond p \rightarrow \Box\Diamond p$ is valid in $\langle W, R \rangle$ iff

$$\forall S \subseteq W \forall w \in W(\exists x(wRx \; \& \; x \in S) \\ \rightarrow \forall x(wRx \rightarrow \exists y(xRy \; \& \; y \in S))).$$

A relation E is called *euclidean* if for all a, b, c, if aEb and aEc, then bEc. And the right-hand side of our last biconditional holds if and only if R is euclidean. (In the right–left direction, suppose aRb and aRc, and take $S = \{c\}$.) So $\Diamond p \rightarrow \Box\Diamond p$ is valid in $\langle W, R \rangle$ iff R is a euclidean relation on $\langle W, R \rangle$.

The notion of a euclidean relation may not be familiar, but it is easy to see that a relation is reflexive and euclidean if and only if it is an equivalence relation, that is, if and only if it is reflexive, symmetric, and transitive. For if E is reflexive and euclidean, E is symmetric (if aEb, then aEb and aEa, whence bEa) and transitive (if aEb and bEc, then bEa by symmetry and bEc, whence aEc); conversely, if E is symmetric and transitive, then E is euclidean (if aEb and aEc, then bEa, and so bEc).

We have noted that all theorems of K are valid in all models. Since $\Box p \rightarrow p$ is valid in a frame $\langle W, R \rangle$ iff R is reflexive, every substitution instance of $\Box p \rightarrow p$, that is, every sentence $\Box A \rightarrow A$, is valid in all reflexive frames, and hence in all reflexive models. Likewise, every sentence $\Box A \rightarrow \Box\Box A$ ($A \rightarrow \Box\Diamond A$, $\Diamond A \rightarrow \Box\Diamond A$) is valid in all transitive (symmetric, euclidean) models. We have thus also established *soundness* theorems for the six systems K, T, $K4$, $S4$, B, and $S5$: Every theorem of K (T, $K4$, $S4$, B, $S5$) is valid in all (all reflexive, all transitive, all reflexive and transitive, all reflexive and symmetric, all reflexive and euclidean) models. We shall call a frame or a model *appropriate to* T ($K4$, $S4$, B, $S5$) if it is reflexive (transitive, reflexive and transitive, reflexive and symmetric, reflexive and euclidean). *Every* frame (model) is *appropriate to K*. In the next chapter we shall prove the *completeness* theorems that are the converses of our six soundness theorems; we shall show that a modal sentence is a theorem of one of the six systems if it is valid in all frames (models) appropriate to that system.

We come now to G. We have seen that each of the four particular modal sentences that we have considered is valid in a frame if and only if the frame satisfied a certain condition, which could be "read off" directly from the sentence, and in each of the four cases we were able to see how to formulate a much simpler, but equivalent, condition. In fact, given any modal sentence, we can always find some "second-order" condition that is satisfied by a frame if and only if the sentence is valid in the frame. The condition will be expressed by a sentence beginning with as many universal quantifiers ranging over subsets of the domain of the frame as there are sentence letters in the sentence; after these will come a universal first-order quantifier ranging over members of the domain; and then will come a formula containing only first-order quantifiers, a binary relation letter, set variables, and propositional connectives. But, although we can always find some such second-order condition, we cannot always hope to recast it as a first-order condition on the accessibility relation, as we were able to for the four second-order conditions on frames corresponding to $\Box p \to p$, $\Box p \to \Box\Box p$, $p \to \Box\Diamond p$, and $\Diamond p \to \Box\Diamond p$.

A case in point is the condition corresponding to $\Box(\Box p \to p) \to \Box p$. We can directly see that $\Box(\Box p \to p) \to \Box p$ is valid in a frame $\langle W, R\rangle$ if and only if $\forall S \subseteq W \forall w \in W(\forall x(wRx \to (\forall y(xRy \to y \in S) \to x \in S)) \to \forall x(wRx \to x \in S))$. Let us call a frame satisfying this condition *appropriate to* G. Can we give an illuminating answer to the question, 'When is a frame appropriate to G?'

We can. We shall see that $\langle W, R\rangle$ is appropriate to G if and only if R is a transitive relation on W and there is no infinite sequence $w_0, w_1, w_2, \ldots$ such that $w_0Rw_1Rw_2 \cdots$.

A relation E is said to be *well-founded* if every nonempty set contains a member to which no member of the set bears E.

The basic fact about well-founded relations
E is well-founded if and only if there is no infinite sequence $a_0, a_1, a_2, \ldots$ such that $\cdots a_2Ea_1Ea_0$.

Proof. Suppose that S is nonempty, but for every a in S there is a b in S such that bEa. Since S is nonempty, for some a_0, $a_0 \in S$. So for some a_1 in S, a_1Ea_0. So for some a_2 in S, a_2Ea_1. So for some $a_3 \ldots$. So for some infinite sequence $a_0, a_1, a_2, \ldots, \cdots a_2Ea_1Ea_0$. [A formalization of this argument in set theory would appeal to the axiom of (dependent) choice.] Conversely, if $\cdots a_2Ea_1Ea_0$, then the set $\{a_0, a_1, a_2, \ldots\}$ is nonempty, but for every i, $a_{i+1}Ea_i$. ⊣

$\breve{R}$ is the converse of the relation R: $a\breve{R}b$ iff bRa.

Theorem 2
A frame $\langle W, R\rangle$ is appropriate to G if and only if R is transitive and $\breve{R}$ is well-founded.

Proof. Suppose that $\langle W, R\rangle$ is appropriate to G. Then for every subset S of W and every w in W, if $\forall x(wRx \rightarrow \forall y(xRy \rightarrow y \in S) \rightarrow x \in S)$, then $\forall x(wRx \rightarrow x \in S)$. We first show that R is a transitive relation on W. For w in W, let $S_w = \{a|wRa\} \cap \{a|\forall b(aRb \rightarrow wRb)\}$. If wRx and $\forall y(xRy \rightarrow y \in S_w)$, then $x \in S_w$: for wRx, and if xRb, then $b \in S_w$, and so wRb. Substituting 'S_w' for 'S', we infer that for every x such that wRx, $x \in S_w$ and therefore for every x such that wRx, for every b such that xRb, wRb; that is, for every x and every b, if wRx and xRb, then wRb. But w was an arbitrary member of W, and R is therefore a transitive relation on W.

We now show that $\breve{R}$ is well-founded: $\breve{R}$ is well-founded iff every nonempty set contains a member that bears R to no member of the set. So suppose that S' is nonempty, but every member of S' bears R to some member of S'. Let $w \in S'$. Then for some z in S', wRz. Let $S = W - S'$. For all x, if wRx and $\forall y(xRy \rightarrow y \in S)$, then $x \in S$ (otherwise for some y, xRy and $y \in S'$). But then, since $\langle W, R\rangle$ is appropriate to G, $\forall x(wRx \rightarrow x \in S)$, which is impossible, since $z \in S'$.

For the converse, suppose that R is transitive, $\breve{R}$ is well-founded, $S \subseteq W$, $w \in W$, wRx', and $x' \notin S$. We must find an x such that wRx, $\forall y(xRy \rightarrow y \in S)$, and $x \notin S$. Let $S' = (W - S) \cap \{z|wRz\}$. Since $x' \in S'$, S' contains a member x that bears R to no member of S'. wRx. $x \notin S$. Suppose that xRy. Then $y \notin S'$. As R is transitive, wRy. But then $y \notin W - S$, and so $y \in S$. ⊣

We shall say that a relation R is *well-capped* if $\breve{R}$ is well-founded. A relation R is well-capped if and only if there is no infinite sequence $w_0, w_1, w_2, \ldots$ such that $w_0Rw_1Rw_2 \cdots$. Well-capped relations are irreflexive: For if wRw, then $wRwRw \cdots$. Relations that are transitive and well-founded are sometimes called *strict partial well-orderings*. Since a relation is transitive if and only if its converse is transitive, R is appropriate to G if and only if $\breve{R}$ is a strict partial well-ordering.

The statement that R is a transitive relation on W and there is no sequence $w_0, w_1, w_2, \ldots$ such that $w_0Rw_1Rw_2 \cdots$ is, to say the least, rather more perspicuous than the statement that $\forall S \subseteq W \forall w \in W$ $(\forall x(wRx \rightarrow (\forall y(xRy \rightarrow y \in S) \rightarrow x \in S))$ $\rightarrow \forall x(wRx \rightarrow x \in S))$. But there is, however, no first-order condition that is satisfied by just those frames that are appropriate to G: Let us say that a modal sentence *characterizes* a class of frames if and only if it is valid in all and only the frames in the class. A sentence of the first- or second-order predicate calculus containing just one binary predicate letter ρ may be said to *describe* a class of frames if it is true in all and only the frames in the class. (A frame may be regarded as a structure for a language whose sole nonlogical symbol is the binary predicate letter ρ.) $\Box p \to p$ characterizes the class of frames described by the first-order sentence $\forall w\ w\rho w$; $\Box p \to \Box\Box p$, those described by $\forall w \forall x \forall y(w\rho x \ \&\ x\rho y \to w\rho y)$; $p \to \Box\Diamond p$, those described by $\forall w \forall x(w\rho x \to x\rho w)$; and $\Diamond p \to \Box\Diamond p$, those described by $\forall w \forall x \forall y(w\rho x \ \&\ w\rho y \to x\rho y)$. We want to see that there is *no first-order* sentence that describes the class of transitive, well-capped frames, the class characterized by $\Box(\Box p \to p) \to \Box p$.

We can show this by making use of the compactness theorem for first-order logic. For suppose there were such a sentence σ. Let $\alpha_0, \alpha_1, \ldots$ be a sequence of distinct individual constants. Then every finite subset of $\{\sigma\} \cup \{\alpha_i \rho \alpha_j | i < j\}$ has a model, and by the compactness theorem the entire set has a model $\mathcal{A}, = \langle A, R, a_0, a_1, \ldots \rangle$. But the binary relation R that interprets ρ in $\mathcal{A}$ cannot be well-capped (because $a_0 R a_1 R a_2 \cdots$), and thus the frame $\langle A, R \rangle$ is not well-capped either, even though σ is true in $\mathcal{A}$ and therefore also true in $\langle A, R \rangle$.

Theorem 2 gives us a soundness theorem for G: All theorems of G are valid in all transitive well-capped frames, and hence in all transitive well-capped models, for the axioms of G that are not axioms of K are just the substitution instances of $\Box(\Box p \to p) \to \Box p$. In Chapter 7, we shall prove that the theorems of G are exactly those modal sentences that are valid in all transitive, well-capped models. Well-founded relations have a well-known and very attractive property: Proofs by induction on well-founded relations are possible. In Chapter 9 we shall take advantage of the opportunity of giving a proof by induction on the converse of the accessibility relation of frames appropriate to G. We shall infer by induction that all sentences of a certain sort hold at all worlds in all models appropriate to G, then deduce by the completeness theorem for G that these sentences are theorems of G, and then apply Theorem 3 of Chapter 3 to prove something quite interesting about PA, namely, that there is a quite simple procedure for calculating from each predicate $(A, \phi)(x)$, where A is modalized in p and contains no sentence letters but p, a deictic sentence that is equivalent to every fixed point of the predicate.

In Chapter 4 we showed that for every letterless sentence A there is a truth-functional combination B of lies such that $\vdash_G A \leftrightarrow B$. We conclude by using our soundness theorem for G to show that this result about letterless sentences is not generalizable to arbitrary modal sentences. We shall show that if B is a truth-functional combination of sentences of the forms $\Box^m p$, $\Box^n - p$, $m, n \geqslant 0$, then it is *not* the case that $\vdash_G B \leftrightarrow \Box(p \vee \Box p)$. (Solovay has established much stronger results along these lines.[2])

To see this, consider the two models:

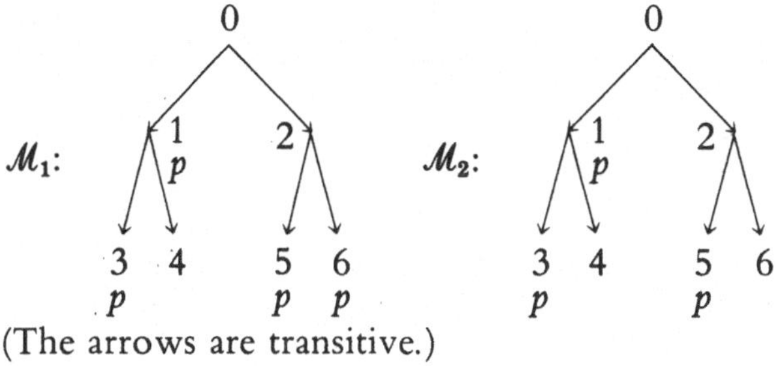

(The arrows are transitive.)

(That is, $W_1 = W_2 = \{0, 1, \ldots, 6\}$; $R_1 = R_2 = \{\langle 0, 1\rangle, \ldots, \langle 0,6\rangle, \langle 1,3\rangle, \langle 1,4\rangle, \langle 2,5\rangle, \langle 2,6\rangle\}$; $P_1(1,p) = P_2(1,p) = P_1(3,p) = P_2(3,p) = P_1(5,p) = P_2(5,p) = P_1(6,p) = \top$; P_1 and $P_2 = \bot$ elsewhere; $\mathcal{M}_1 = \langle W_1, R_1, P_1\rangle$; $\mathcal{M}_2 = \langle W_2, R_2, P_2\rangle$.) $\mathcal{M}_1$ and $\mathcal{M}_2$ are appropriate to G. Note that $\mathcal{M}_1 \vDash_0 \Box^m p$ iff $m \geqslant 3$; $\mathcal{M}_2 \vDash_0 \Box^m p$ iff $m \geqslant 3$; $\mathcal{M}_1 \vDash_0 \Box^n - p$ iff $n \geqslant 3$; and $\mathcal{M}_2 \vDash_0 \Box^n - p$ iff $n \geqslant 3$. Thus $\mathcal{M}_1 \vDash_0 B$ iff $\mathcal{M}_2 \vDash_0 B$ for all truth-functional combinations B of sentences of the forms $\Box^m p$, $\Box^n - p$, $m, n \geqslant 0$. But $\mathcal{M}_1 \vDash_0 \Box(p \vee \Box p)$ and $\mathcal{M}_2 \nvDash_0 \Box(p \vee \Box p)$. Thus if $\vdash_G B \leftrightarrow \Box(p \vee \Box p)$, then $\mathcal{M} \vDash_0 B \leftrightarrow \Box(p \vee \Box p)$ and $\mathcal{M}_2 \vDash_0 B \leftrightarrow \Box(p \vee \Box p)$, whence $\mathcal{M}_1 \vDash_0 B$ and $\mathcal{M}_2 \nvDash_0 B$, which is absurd.

Exercises

1 Define nest(A) by nest($\bot$) = nest(p) = 0; nest($A \rightarrow B$) = max(nest(A), nest(B)); nest($\Box A$) = nest(A) + 1. True or false: If A contains at most one sentence letter and $\nvdash_G - A$, then, for some W, R, P, w, $\{W,R,P\} \vDash_w A$ and for all w_0, w_1, . . . , $w_{\text{nest}(A)}$ in W, not: $w_0 R w_1 R \cdots R w_{\text{nest}(A)}$.

2 Show that no sentence is valid in all and only those frames that are well-capped. [*Hint:* Let S be the successor relation on the set N of natural numbers. Suppose that $\langle N,S,P\rangle \nvDash_m A$. Let $W = \{0,1,\ldots,\text{nest}(A)\}$, and let R be the successor relation on W. Let $Q(r,p) = P(m + r,p)$ for all $r \leqslant \text{nest}(A)$. Show that for every subsentence B of A and every $r \leqslant \text{nest}(A) - \text{nest}(B)$, $\langle N,S,P\rangle \vDash_{m+r} B$ iff $\langle W,R,Q\rangle \vDash_r B$.]

6

Canonical models

We shall now present a method, due to Scott and Makinson,[1] for constructing modal-logical models. The method enables us to construct from each consistent normal modal propositional logic L a model $\mathcal{M}_L$, called *the canonical model for L,* in which all and only the theorems of the logic are valid. Canonical models were introduced to prove the completeness, with respect to the appropriate semantics, of a number of systems of modal logic. In the present chapter we shall use them to give proofs of completeness of the six systems K, $K4$, T, $S4$, B, and $S5$, and in the next chapter we shall show how to make use of $\mathcal{M}_G$, the canonical model for G, to prove the completeness of G. In Chapter 11, $\mathcal{M}_G$ will figure prominently in the proof of the de Jongh–Sambin Fixed-Point Theorem.

'Complete' (said of a system of logic) may mean various things in various contexts, but almost always, when a system is called complete, part of the meaning is that *only* the theorems of the system have some noteworthy property, for example, validity in all models in a certain class. In the present chapter, when we say that a system of modal logic is complete, we shall mean that it is only the theorems of the system that are valid in all models appropriate to the system. (In Chapter 12, where we prove Solovay's Completeness Theorem for G, 'completeness' will refer to the provability in G of all modal sentences all of whose translations are theorems of PA.)

In the last chapter we saw that if a sentence A is a theorem of K, $K4$, T, $S4$, B, or $S5$, then A is valid in all models, all transitive models, all reflexive models, all reflexive and transitive models, all reflexive and symmetric models, or all reflexive and euclidean models, respectively. We now want to demonstrate the converses of all six of these assertions. Our strategy will be to define the canonical model $\mathcal{M}_L$ for a consistent normal modal logic L, that is, a normal modal logic of which $\perp$ is not a theorem, and then show that all and only the theorems of L are valid in $\mathcal{M}_L$. The completeness of K will thus be established, for if a modal sentence is valid in all models, then it is valid in $\mathcal{M}_K$, and hence is a theorem of K. We shall then show $\mathcal{M}_{K4}$ to be transitive, $\mathcal{M}_T$ reflexive, $\mathcal{M}_{S4}$ reflexive and transitive, $\mathcal{M}_B$ reflexive and symmetric, $\mathcal{M}_{S5}$ reflexive and euclidean. The appropriate completeness theorem for each of these five systems will then have been established, for a modal sentence will then have been shown to be a theorem of $K4$, etc., if it is valid in all transitive models, etc. Along the way, we shall notice that $\mathcal{M}_G$ is transitive. Unfortunately, $\mathcal{M}_G$ is not well-capped, as we shall see at the end of the chapter, and we cannot prove that A is a theorem of G if A is valid in all transitive and well-capped models simply by adverting to the relation-theoretical character of the accessibility relation of the canonical model for G. The additional work that is needed is done in Chapter 7.

[Let us note that in order to establish the completeness of, e.g., $K4$, it is necessary only to show that if A is not a theorem of $K4$, then there is a transitive model (which may depend on A) in which A is not valid. It is not necessary to show that there is a single transitive model in which all nontheorems of $K4$ are invalid. However, this stronger assertion is in fact true, and we shall prove the completeness of $K4$ by proving it.]

How does one go about proving that for every consistent normal modal logic there is a model in which all and only the theorems of the logic are valid? The idea is to combine Kripke's notion of accessibility (relative possibility), Carnap's explication of a possible world as a set of sentences of a certain kind, and Lindenbaum's theorem that any consistent theory can be extended to a complete and consistent theory in the following manner: Fix a normal modal logic L. Call a set S of modal sentences *maximal* (L-)*consistent* if S is consistent and contains each modal sentence or its negation. Following Carnap, let the worlds of the model under construction be the maximal consistent sets of sentences, with a view to showing a sentence true at a world if and only if it is a member of the maximal consistent set that that world is. Following Kripke, let a world x be accessible from a world w if, for all modal sentences A, A is a member of x whenever $\Box A$ is a member of w – equivalently, if $\Diamond A$ is a member of w whenever A is a member of x. Then the choice of the evaluator is forced by the wish to show sentences true at just those worlds to which they belong: The evaluator must be the function that assigns to each pair consisting of a world and a sentence letter the value $\top$ iff the letter belongs to the world. Then, following Lindenbaum, show that if A is not a theorem of L, then, since $\{-A\}$ is consistent, there is a maximal consistent set to which $-A$ belongs. Finally, following Scott and Makinson, show that a world is everything that is the case, that is, that *truth at* a world always coincides with *belonging to* that world, and conclude that if A is not a theorem, then $-A$ is true at, and hence A is false at, some world of the model.

The deep part of the proof is, of course, the demonstration that A is true at a world w if and only if A is in w. It is easy enough to see that the assertion holds if A is a sentence letter or, as we are dealing with maximal consistent sets, a truth-functional compound of sentences for which the assertion holds. And then, if A, $= \Box B$, is in w

and x is accessible from w, then B is in x (by the definition of accessibility), and so, by the hypothesis of the induction we may take ourselves to be making, B is true at x; thus if $\Box B$ is in w, $\Box B$ is true at w. And if $\Box B$ is true at w, then for all x accessible from w, B is true at x, and hence by the inductive hypothesis, B is in x. But here is the question we need to answer in order to complete the proof: How do we know that $\Box B$ is in w whenever B is in all x accessible from w? The answer is contained in the proof of Lemma 3.

The task then remains of showing, where possible, that the accessibility relation of $\mathcal{M}_L$ has the right relation-theoretical character (i.e., is transitive if L is $K4$, etc.), but as we shall see, this task is often quite easy.

We now begin the proofs.

Assume that L is a consistent normal modal logic. A set of sentences is called *L-consistent* if there is no conjunction of members of the set whose negation is a theorem of L. Thus S is L-consistent if for every $A_1, \ldots, A_n$ in S, not $\vdash_L -(A_1 \& \cdots \& A_n)$. As in Chapter 4, we identify the empty conjunction with $\top$; thus since L is consistent, the null set is L-consistent. Let us notice that the definition of L-consistency does not mention models at all: S is L-consistent just in case there is no proof in L whose last sentence is the negation of a conjunction of members of S. (In what follows, we frequently omit 'L-'.)

Here is a lemma about consistency.

Lemma 1
If S is consistent, then either $S \cup \{A\}$ is consistent or $S \cup \{-A\}$ is consistent (for every modal sentence A).

Proof. If $S \cup \{A\}$ is inconsistent, then there is a conjunction of members of $S \cup \{A\}$ whose negation is a theorem of L and therefore, for some $B_1, \ldots, B_j$ in S,

$\vdash_L A \to -(B_1 \,\&\, \cdots \,\&\, B_j)$; similarly, if $S \cup \{-A\}$ is inconsistent, then, for some $C_1, \ldots, C_k$ in S, $\vdash_L -A \to -(C_1 \,\&\, \cdots \,\&\, C_k)$. Consequently, if both $S \cup \{A\}$ and $S \cup \{-A\}$ are inconsistent, then $\vdash_L -(B_1 \,\&\, \cdots \,\&\, B_j) \vee -(C_1 \,\&\, \cdots \,\&\, C_k)$, and so $\vdash_L -(B_1 \,\&\, \cdots \,\&\, B_j \,\&\, C_1 \,\&\, \cdots \,\&\, C_k)$. But then, since all the Bs and Cs are in S, S is inconsistent. ⊣

A set of sentences is called *maximal (L-)consistent* ("maximal," for short) if it is consistent and either contains A or contains $-A$ for every modal sentence A. If S is a maximal set (or even a consistent set), then it is never the case that both A and $-A$ are in S, for, certainly, $\vdash_L -(A \,\&\, -A)$. Thus *if S is maximal, then exactly one of A and $-A$ is in S.*

(It follows that if S is maximal, every set of sentences of which it is a proper subset is inconsistent, for if $S \subset T$, A is in T; but A is not in S, then $-A$ is in S and hence both A and $-A$ are in T. Conversely, if S is consistent, but not maximal, then for some sentence A, neither A nor $-A$ is in S and therefore S is a proper subset of some consistent set of sentences, for by Lemma 1, at least one of $S \cup \{A\}$ and $S \cup \{-A\}$ is consistent.)

Lemma 2

Every consistent set is a subset of some maximal set.

Proof. Suppose that S is a consistent set. Let $B_0, B_1, B_2, \ldots$ be an enumeration of all modal sentences. We define a sequence $S_0, S_1, S_2, \ldots$ of sets of sentences by putting $S_0 = S$;

$$S_{i+1} = \begin{cases} S_i \cup \{B_i\} \text{ if } S_i \cup \{B_i\} \text{ is consistent,} \\ S_i \cup \{-B_i\} \text{ otherwise.} \end{cases}$$

It is evident that if $i \leq j$, $S_i \subseteq S_j$.

Every S_i is consistent. For S_0, $= S$, is consistent by our assumption. Suppose that S_i is consistent. Then if $S_i \cup \{B_i\}$ is consistent, $S_{i+1} = S_i \cup \{B_i\}$; but if $S_i \cup \{B_i\}$ is not consistent, then by Lemma 1, $S_i \cup \{-B_i\}$ is consistent, and, by the definition of S_{i+1}, $S_{i+1} = S_i \cup \{-B_i\}$. In either case S_{i+1} is consistent.

Let T be the union of all the S_is; that is, a sentence is in T if and only if it is in some S_i. $S = S_0 \subseteq S_i \subseteq T$.

T is consistent. For if not, then, for some $A_1, \ldots, A_n$ in T, $\vdash_L -(A_1 \& \cdots \& A_n)$. Since A_1 is in T, A_1 is in some S_{i_1}; . . . ; since A_n is in T, A_n is in some S_{i_n}. Let $i = \max(i_1, \ldots, i_n)$. Then all of $A_1, \ldots, A_n$ are in S_i, which cannot then be consistent, as $\vdash_L -(A_1 \& \cdots \& A_n)$. But we have just seen that S_i is consistent. The contradiction shows that T is consistent.

T is maximal. For if B_i is not in T, then B_i is not in S_{i+1}, which is a subset of T, and so $S_i \cup \{B_i\}$ is inconsistent, so $S_{i+1} = S_i \cup \{-B_i\}$, and so $-B_i$ is in T, which includes S_{i+1}. Thus if a sentence is not in T, then its negation is in T, and therefore S is a subset of the maximal set T. $\dashv$

There are two more facts about maximal consistent sets that we should notice before giving the official definition of $\mathcal{M}_L$.

Every maximal consistent set contains every theorem of L. For if S is maximal, $\vdash_L A$, and $A \notin S$, then $-A \in S$. But $\vdash_L --A$, and there is thus a conjunction whose sole conjunct is a member of S and whose negation is a theorem of L.

Every maximal set is closed under modus ponens. For if S is maximal and both A and $(A \rightarrow B)$ are in S, then B is also in S, for otherwise $-B$ is in S and consequently $(A \& (A \rightarrow B) \& -B)$ is a conjunction of members of S, whose negation, which is a tautology, is a theorem of L.

Let W_L be the set of all maximal L-consistent sets of sentences. (Since L is consistent, W_L is not empty.)

Let $R_L = \{\langle w,x\rangle | w,x \in W_L$ and for all modal sentences A, if $\Box A \in w$, then $A \in x\}$. So if w and x are in W_L, then wR_Lx iff $\{A|\Box A \in w\} \subseteq x$.

Let P_L be the function that assigns to each pair consisting of a member w of W_L and a sentence letter p, the truth-value $\top$ iff $p \in w$, (and assigns $\bot$ to the pair otherwise).

Let $\mathcal{M}_L = \langle W_L, R_L, P_L\rangle$.

Lemma 3

Suppose that $w \in W_L$ and for all x such that wR_Lx, $B \in x$. Then $\Box B \in w$.

Proof. Let $Y = \{A|\Box A \in w\}$. If $Y \cup \{-B\}$ is consistent, then by Lemma 2 there is a maximal set x such that $Y \cup \{-B\} \subseteq x$. Since $Y \subseteq x$, wR_Lx. But since $-B \in x$, $B \notin x$, which contradicts the supposition of the lemma. Consequently, $Y \cup \{-B\}$ is not consistent, and therefore there are $A_1, \ldots, A_n$ in Y such that $\vdash_L -B \rightarrow -(A_1 \& \cdots \& A_n)$. Then $\vdash_L (A_1 \& \cdots \& A_n) \rightarrow B$, and so, since L is normal, $\vdash_L \Box(A_1 \& \cdots \& A_n) \rightarrow \Box B$. But since $\vdash_L \Box(A_1 \& \cdots \& A_n) \leftrightarrow (\Box A_1 \& \cdots \& \Box A_n)$ (normality again), $\vdash_L \Box A_1 \& \cdots \& \Box A_n \rightarrow \Box B$, and so $\vdash_L \Box A_1 \rightarrow (\Box A_2 \rightarrow (\cdots (\Box A_n \rightarrow \Box B) \cdots))$.

Since every maximal set contains every theorem of L, $\Box A_1 \rightarrow (\Box A_2 \rightarrow (\cdots (\Box A_n \rightarrow \Box B) \cdots))$ is in w. And since $A_1, \ldots, A_n$ are all in Y, $\Box A_1, \ldots, \Box A_n$ are all in w (cf. the definition of Y). And because every maximal set is closed under modus ponens, $\Box B$ is also in w. $\dashv$

Here is the fundamental theorem about canonical models.

Theorem 1

For every modal sentence A and every w in W_L, $\mathcal{M}_L \vDash_w A$ iff $A \in w$.

Proof. Induction on the complexity of A.

If A is $\bot$, then since $\vdash_L -\bot$, $\bot \notin w$. But $\mathcal{M}_L \nvDash_w \bot$ also.

If A is a sentence letter p, then $\mathcal{M}_L \vDash_w p$ iff $P_L(w,p) = \top$ iff $p \in w$.

If A is $(B \to C)$, then $\mathcal{M}_L \vDash_w (B \to C)$ iff either $\mathcal{M}_L \nvDash_w B$ or $\mathcal{M}_L \vDash_w C$, iff (by the induction hypothesis) either $B \notin w$ or $C \in w$, iff (by the maximality of w) either $-B \in w$ or $C \in w$. If either $-B \in w$ or $C \in w$, then since both $\vdash_L -B \to (B \to C)$ and $\vdash_L C \to (B \to C)$, $(B \to C) \in w$. But if both $-B \notin w$ and $C \notin w$, then by the maximality of w, $B \in w$ and $-C \in w$, and so, since $\vdash_L B \to (-C \to -(B \to C))$, $-(B \to C) \in w$, whence $(B \to C) \notin w$.

Lastly, if A is $\Box B$, then if $\mathcal{M}_L \vDash_w \Box B$, for every x such that wR_Lx, $\mathcal{M}_L \vDash_x B$, and then, by the induction hypothesis, for every x such that wR_Lx, $B \in x$. By Lemma 3, $\Box B \in w$. Conversely, if $\Box B \in w$, then by the definition of R_L, for all x such that wR_Lx, $B \in x$; by the induction hypothesis, for all x such that wR_Lx, $\mathcal{M}_L \vDash_x B$, whence $\mathcal{M}_L \vDash_w \Box B$. ⊣

Theorem 2

$\vdash_L A$ iff A is valid in $\mathcal{M}_L$.

Proof. If $\vdash_L A$, then A is in every w in W_L, and therefore by Theorem 1 for every w in W_L, $\mathcal{M}_L \vDash_w A$, that is, A is valid in $\mathcal{M}_L$. Conversely, if $\nvdash_L A$, then $\{-A\}$ is L-consistent. By Lemma 2, for some maximal set w, $-A \in w$, and thus by Theorem 1, for some w in W_L, $\mathcal{M}_L \vDash_w -A$, and therefore A is not valid in $\mathcal{M}_L$. ⊣

We have thus established the first of our six completeness theorems: If A is valid in all models, then $\vdash_K A$. (For if A is valid in all models, then A is valid in $\mathcal{M}_K$ in particular, and then by Theorem 2, $\vdash_K A$.) The other five follow quite directly from the next theorem.

Theorem 3
If every sentence of the form $\Box A \to \Box\Box A$ ($\Box A \to A$, $A \to \Box\Diamond A$, $\Diamond A \to \Box\Diamond A$) is a theorem of L, then $\mathcal{M}_L$ is transitive (reflexive, symmetric, euclidean).

Proof
$\Box A \to \Box\Box A$: We must show that for all w, x, y in W_L, if wR_Lx and xR_Ly, then wR_Ly. Assume wR_Lx and xR_Ly, that is, for every B, if $\Box B \in w$, then $B \in x$, and for every C, if $\Box C \in x$, then $C \in y$. We must show that for every A, if $\Box A \in w$, then $A \in y$. Suppose $\Box A \in w$. Then, since $\vdash_L \Box A \to \Box\Box A$, $\Box\Box A \in w$. But then $\Box A \in x$, and so $A \in y$.

$\Box A \to A$: We must show that wR_Lw. But since $\vdash_L \Box A \to A$, if $\Box A \in w$, then $A \in w$.

$A \to \Box\Diamond A$: Suppose that wR_Lx. We must show that if $\Box B \in x$, then $B \in w$. Suppose $B \notin w$. Then $-B \in w$. Since $\vdash_L -B \to \Box\Diamond -B$, $\Box\Diamond -B \in w$, so $\Diamond -B \in x$, so $-\Box B \in x$, and so $\Box B \notin x$.

$\Diamond A \to \Box\Diamond A$: Suppose that wR_Lx and wR_Ly. We must show that if $\Box B \in x$, then $B \in y$. Suppose $B \notin y$. Then $\Box B \notin w$, $-\Box B \in w$, and so $\Diamond -B \in w$. Since $\vdash_L \Diamond -B \to \Box\Diamond -B$, $\Box\Diamond -B \in w$, $\Diamond -B \in x$, $-\Box B \in x$, and $\Box B \notin x$. ⊣

Since every sentence of the form $\Box A \to \Box\Box A$ is a theorem of G, it follows from Theorem 3 that $\mathcal{M}_G$ is transitive.

Theorem 4

(a) $\vdash_K A$ iff A is valid in all models;
(b) $\vdash_{K4} A$ iff A is valid in all transitive models;
(c) $\vdash_T A$ iff A is valid in all reflexive models;

(d) $\vdash_{S4} A$ iff A is valid in all reflexive and transitive models;
(e) $\vdash_{B} A$ iff A is valid in all reflexive and symmetric models;
(f) $\vdash_{S5} A$ iff A is valid in all reflexive and euclidean models.

Proof. The "only if" parts were shown in Chapter 5 and (a) has already been established. By Theorem 3, $\mathcal{M}_{K4}$ is transitive, $\mathcal{M}_T$ is reflexive, $\mathcal{M}_{S4}$ is reflexive and transitive, $\mathcal{M}_B$ is reflexive and symmetric, and $\mathcal{M}_{S5}$ is reflexive and euclidean. Then if A is valid in all transitive models, A is valid in $\mathcal{M}_{K4}$, which is transitive, and then by Theorem 2, A is a theorem of $K4$. The proofs for T, $S4$, B, and $S5$ are similar. $\dashv$

The techniques of this chapter yield an alternative proof of the result, demonstrated in Chapter 1, that a normal system extends $S5$ if and only if it extends $S4$ and extends B: A relation is reflexive and euclidean if and only if it is reflexive, symmetric, and transitive (as we noted in the last chapter). Thus if L is a normal system that extends $S5$ (that extends both $S4$ and B), then $\mathcal{M}_L$ is reflexive, symmetric, and transitive, and therefore all theorems of $S4$ and all theorems of B (all theorems of $S5$) are valid in $\mathcal{M}_L$, and hence by Theorem 2, L extends both $S4$ and B (extends $S5$).

We remarked earlier that $\mathcal{M}_G$ is not well-capped. This may be shown fairly easily if we make use of Theorem 1 of Chapter 4, according to which for every letterless sentence A there is a truth-functional combination B of lies such that $\vdash_G A \leftrightarrow B$.

Let S be the set containing all sentence letters, all necessitations of sentence letters, and all negations of lies. Then if ϕ is the realization that assigns $\top$ to all sentence letters, and C is a conjunction of members of S, then C^{ϕ} is true: There is therefore no conjunction of members of S whose negation is a theorem of G. S is therefore G-consistent.

We shall write '$S \vdash_G A$' to mean '$S \cup \{-A\}$ is not G-consistent'. Since S contains all sentence letters and their necessitations, we have $S \vdash_G p \leftrightarrow \top$ and $S \vdash_G \Box(p \leftrightarrow \top)$ for every sentence letter p. For each modal sentence B, let $B^\top$ be the result of replacing each sentence letter in B by $\top$. An easy induction on the complexity of B shows that both $S \vdash_G B \leftrightarrow B^\top$ and $S \vdash_G \Box(B \leftrightarrow B^\top)$ for all sentences B. [In the case in which $B = \Box C$, we appeal to the fact that $\vdash_G \Box(C \leftrightarrow C^\top) \rightarrow \Box\Box(C \leftrightarrow C^\top)$.] Furthermore, since S contains all negations of lies, and every letterless sentence is equivalent in G to a truth-functional combination of lies, we have that, for every sentence B, either $S \vdash_G B$ or $S \vdash_G -B$. Thus S has *exactly one* maximal consistent extension w, which contains just those sentences B such that $S \vdash_G B$. And since all negations of lies are true, if C is letterless, then $S \vdash_G C$ if and only if C^Φ is true.

Suppose now that $\Box A \in w$. Then $S \vdash_G A \leftrightarrow A^\top$, $S \vdash_G \Box(A \leftrightarrow A^\top)$, $S \vdash_G \Box A \leftrightarrow \Box A^\top$, $(\Box A \leftrightarrow \Box A^\top) \in w$, $(A \leftrightarrow A^\top) \in w$, $\Box A^\top \in w$, $S \vdash_G \Box A^\top$, $(\Box A^\top)^\Phi$ is true, $\vdash_{PA} (A^\top)^\Phi$, $(A^\top)^\Phi$ is true, $S \vdash_G A^\top$, $A^\top \in w$, and $A \in w$. So if $\Box A \in w$, then $A \in w$, for every sentence A. We thus have that wR_Gw, whence $wR_GwR_Gw \cdots$, and therefore $\mathcal{M}_G$ is not well-capped.

Exercise (Solovay)
We have supposed that there is a countably infinite sequence of sentence letters, and it is therefore evident that there are exactly $2^{\aleph_0}$ maximal G-consistent sets of sentences. Let us say that $T \approx_p T'$ if T and T' are maximal G-consistent sets of sentences such that every sentence containing no letter other than p is in T iff in T'. Show that there are exactly $2^{\aleph_0}$ $\approx_p$-equivalence classes. (*Hint:* If $X \subseteq \omega$, let $S_X = \{\Diamond^i\top : i \in \omega\} \cup \{\Box(\Diamond^i\top \,\&\, -\Diamond^{i+1}\top \rightarrow p) : i \in X\} \cup \{\Box(\Diamond^i\top \,\&\, -\Diamond^{i+1}\top \rightarrow -p) : i \in \omega - X\}$. Then show that S_X is G-consistent and that if $X, X' \subseteq \omega$ and $X \neq X'$, then $S_X \cup S_{X'}$ is not G-consistent.) Say that $T \approx T'$ if T and T' are maximal G-consistent sets such that every letterless sentence is in T iff in T'. Show that there are exactly $\aleph_0$ $\approx$-equivalence classes.

7

The completeness and decidability of G

In Chapter 5 we saw that every theorem of G is valid in all transitive, well-capped models. We are now going to prove the converse, that if a modal sentence is valid in all transitive, well-capped models, then it is a theorem of G, and also prove that there is an effective method (an algorithm) for deciding whether or not any given modal sentence is a theorem of G. In short, we are going to prove that G is complete (with respect to transitive, well-capped models) and decidable. Both of these results will follow from a better completeness theorem for G, the proof of which is the main concern of the present chapter: A modal sentence is a theorem of G if it is valid in all transitive, well-capped models, whose domain is a *finite* set.

There is a useful alternative characterization of transitive, well-capped models whose domain is finite, namely, models $\langle W, R, P \rangle$ in which W is finite and R is a transitive, irreflexive relation on W. (A relation is irreflexive if nothing bears that relation to itself.) For if W is a finite set and R is a transitive relation on W, then R is well-capped if and only if R is irreflexive: It is clear that if R is well-capped, then R is irreflexive. For if R is not irreflexive, then for some w, wRw, and so $wRwRwRw \cdots$, and therefore R is not well-capped. Conversely, suppose that R is not well-capped. Then there is an infinite sequence $w_0, w_1, w_2, \ldots$ of members of W such that $w_0 R w_1 R w_2 \cdots$. Since R is transitive, $w_i R w_j$ whenever $i < j$. And since R is a relation on the finite set W, the sequence repeats at some point, that is, for some i and j, $i < j$ and $w_i = w_j$. But then $w_i R w_i$ and R is not irreflexive.

Frames $\langle W, R \rangle$ in which W is finite and R is a transitive, irreflexive relation on W are called *finite strict partial orderings*. We shall thus show that a modal sentence is a theorem of G if it is valid in all finite strict partial orderings.

The decidability of G follows (in the usual manner) from the fact that a sentence is not a theorem of G if and only if there is a finite model of a certain sort in which it is not valid. To decide upon the theoremhood of any given sentence A, we (effectively) enumerate all the theorems of G and all models $\langle W, R, P \rangle$ in which W is a finite set of integers, R is a transitive and irreflexive relation on W, and P is a function that assigns a truth-value to each pair consisting of a member of W and a sentence letter in A. As theorems appear, we check each one to see whether or not it is identical to A. As models appear, we calculate the truth-value of A in all of the finitely many worlds of each model and determine whether A is false at at least one of them. (The determination can of course be effectively performed.) Either we shall eventually come upon A in the enumeration of theorems, or we shall not; but in the latter case, as our main theorem will show, we shall come upon a model at one of whose worlds A is false, and we shall then know that A is not a theorem of G. (There is obviously no generality lost in restricting attention to finite sets *of integers* and functions on sentence letters *in A*.)

The proof of the main theorem will show that if A is a modal sentence with n subsentences that is not a theorem of G, then there is a finite strict partial ordering in which A is invalid, *whose domain contains* $\leqslant 2^n$ *worlds*.

Thus we can actually specify an upper bound on the number of models that we might have to examine in order to decide the theoremhood of a given sentence. In the next chapter we shall describe a *practical* decision procedure for G.

The proof we now give of the theorem that if $\nvdash_G A$, then A is invalid in some finite strict partial ordering is extracted from Segerberg's *Essay in Classical Modal Logic.*[1] In the next chapter we shall present another proof of the same theorem. Even though Segerberg's proof is perhaps somewhat more difficult to comprehend than that proof, its detailed statement is quite short and it will be used again in Chapter 13.

Recall from Chapter 6 the canonical model $\mathcal{M}_G$ for G. The domain of $\mathcal{M}_G$, W_G, is the set of maximal G-consistent sets of sentences; a world w in W_G bears R_G, the accessibility relation of $\mathcal{M}_G$, to a world x just in case $A \in x$ whenever $\Box A \in w$ (for all sentences A), and $P_G(w,p) = \top$ iff $p \in w$, for all w in W and all sentence letters p. By Theorem 1 of Chapter 6, for every sentence A and every w in W_G, $A \in w$ iff $\mathcal{M}_G \vDash_w A$, and by Theorem 2 of Chapter 6, the theorems of G are precisely the sentences true at all worlds of $\mathcal{M}_G$. We saw in Chapter 6 that $\mathcal{M}_G$ is transitive (i.e., that R_G is).

Suppose now that A is not a theorem of G. Then there is a world t in W_G such that $\mathcal{M}_G \nvDash_t A$. We are going to construct three more models in which A is false; the last of these will have a finite domain and a transitive, irreflexive accessibility relation.

Let $W_1 = \{t\} \cup \{x | tR_Gx\}$.

Let $R_1 = \{\langle w,x\rangle | w,x \in W_1 \ \& \ wR_Gx\}$.

Let $P_1(w,p) = P_G(w,p)$ for all w in W_1 and all sentence letters p. Thus R_1 and P_1 are just the restrictions of R_G and P_G to W_1.

Let $\mathcal{M}_1 = \langle W_1, R_1, P_1\rangle$.

Since R_G is transitive, $R_G = (R_G)_*$ and so by Theorem 1 of Chapter 5, $\mathcal{M}_1 \vDash_x B$ iff $\mathcal{M}_G \vDash_x B$ iff $B \in x$, for every sentence B and every x in W_1. We thus have that $\mathcal{M}_1 \nvDash_t A$.

We are now going to construct a finite model $\mathcal{M}_2$ by "identifying" those worlds of $\mathcal{M}_1$ *to which the same subsentences of A belong.*

If w and x are in W_1, say that $w \equiv x$ if and only if for every subsentence B of A, $B \in w$ iff $B \in x$. $\equiv$ is clearly an equivalence relation on W_1. Let w^o be the $\equiv$-equivalence class of w, that is, $w^o = \{x | w \equiv x\}$.

Let $W_2 = \{w^o | w \in W_1\}$.

Let $R_2 = \{\langle w^o, x^o \rangle |$ for every subsentence $\Box B$ of A, if $\Box B \in w$, then $\Box B \in x$ and $B \in x\}$.

Let $P_2(w^o, p) = P_1(w, p)$ if p is a sentence letter that is a subsentence of A, and (arbitrarily) let $P_2(w^o, p) = \bot$ if not.

Notice that the definitions of R_2 and P_2 are in fact independent of the choice of members of the equivalence classes w^o and x^o.

Let $\mathcal{M}_2 = \langle W_2, R_2, P_2 \rangle$.

W_2 *is finite.* In fact, if there are n subsentences of A, then there are at most 2^n members of W_2.

R_2 *is transitive.* Suppose that wR_2xR_2y. We must show that wR_2y. But if $\Box B$ is a subsentence of A and $\Box B \in w$, then $\Box B \in x$, and so $\Box B \in y$ and $B \in y$.

Observe that if wR_1x, *then* $w^oR_2x^o$: Suppose wR_1x. Then wR_Gx, and thus if $\Box B \in w$, then, since $\vdash_G \Box B \rightarrow \Box\Box B$, we have $\Box\Box B \in w$, so $\Box B \in x$ and $B \in x$, and so $w^oR_2x^o$.

Lemma 1

For every truth-functional combination B of subsentences of A and every w in W_1, $\mathcal{M}_1 \models_w B$ if and only if $\mathcal{M}_2 \models_{w^o} B$.

Proof. Induction on the complexity of B. If p is a sentence letter occurring in A, then $\mathcal{M}_1 \models_{w^o} p$ iff $P_1(w,p) = \top$ iff $P_2(w^o, p) = \top$ iff $\mathcal{M}_2 \models_{w^o} p$. And, clearly, if B is a truth-functional combination of sentences for which the conclusion of the lemma holds, then the lemma also holds for B. We may, therefore, assume that B is a subsentence of A, that $B = \Box C$, and that the lemma holds for C. Suppose that $\mathcal{M}_1 \models_w \Box C$. Then $\Box C \in w$. If $w^o R_2 x^o$, then $C \in x$, $\mathcal{M}_1 \models_x C$, and so, by the induction hypothesis, $\mathcal{M}_2 \models_{x^o} C$. Thus for every x^o such that $w^o R_2 x^o$, $\mathcal{M}_2 \models_{x^o} C$. We thus have that $\mathcal{M}_2 \models_{w^o} \Box C$. Conversely, suppose that $\mathcal{M}_2 \models_{w^o} \Box C$. If $w R_1 x$, then, as we have observed, $w^o R_2 x^o$; and therefore $\mathcal{M}_2 \models_{x^o} C$, and so, by the induction hypothesis, $\mathcal{M}_1 \models_{x^o} C$. Thus whenever $w R_1 x$, $\mathcal{M}_1 \models_x C$, and we have that $\mathcal{M}_1 \models_w \Box C$. ⊣

Since A is a subsentence of itself, A is certainly a truth-functional combination of subsentences of A. And since $\mathcal{M}_1 \not\models_t A$, Lemma 1 implies that $\mathcal{M}_2 \not\models_{t^o} A$. $\mathcal{M}_2$ is thus a transitive model with a finite domain in which A is invalid. However, not all models are irreflexive that are constructed from $\mathcal{M}_G$ and an arbitrary t in W_G in the same manner in which $\mathcal{M}_2$ was constructed from $\mathcal{M}_G$ and a t at which A is false. For, as we saw at the end of Chapter 6, there are worlds w in W_G such that $w R_G w$. And if $\mathcal{M}_2$ is obtained from some such w, then $w R_1 w$ and so $w^o R_2 w^o$ as well, and $\mathcal{M}_2$ is not irreflexive. The difficulty is that there may be nonempty subsets of the field of R_2, each member of which bears R_2 to itself and all other members. We must now show how to "pare" R_2 "down" to a transitive, irreflexive relation R_3 on W_2 in such a way that we can show that for every subsentence B of A and every world w^o in W_2, $\mathcal{M}_2 \models_{w^o} B$ iff $\langle W_2, R_3, P_2\rangle \models_{w^o} B$. But once we have found such an R_3, we shall have shown that there is a finite strict partial ordering, namely, $\langle W_2, R_3\rangle$, in which A is invalid. We now replace R_2 by a transitive, irreflexive *subrelation* R_3 of R_2 as follows:

Say that $w^o \sim x^o$ if and only if either $w^o = x^o$ or both $w^o R_2 x^o$ and $x^o R_2 w^o$. Since R_2 is transitive, $\sim$ is an equivalence relation on W_2.

For each $S \subseteq W_2$ such that for some w, $S = \{y^o | w^o \sim y^o\}$, let L_S be a strict linear ordering of S, that is, a transitive and irreflexive relation on S such that if $x^o, y^o \in S$, either $x^o L_S y^o$ or $x^o = y^o$ or $y^o L_S x^o$. (Of course, if S contains just one member, then L_S is the empty relation.)

Let $R_3 = \{\langle w^o, x^o\rangle | w^o, x^o \in W_2$ and either $w^o \not\sim x^o$ and $w^o R_2 x^o$ or $w^o \sim x^o$ and $w^o L_S x^o$, where $S = \{y^o | w^o \sim y^o\}\}$.

If $w^o \sim x^o$, $w^o \sim y^o$, and $x^o \neq y^o$, then since R_2 is transitive, $x^o R_2 y^o$ and $y^o R_2 x^o$. It follows that R_3 *is a subrelation of* R_2.

Moreover, R_3 *is transitive:* Suppose that $w^o R_3 x^o R_3 y^o$. Then $w^o R_2 x^o R_2 y^o$, and, since R_2 is transitive, $w^o R_2 y^o$. We may thus suppose that $w^o \sim y^o$. Then either $w^o = y^o$ or $y^o R_2 w^o$; in either case, by the transitivity of R_2, $w^o \sim x^o \sim y^o \sim w^o$. But then $w^o L_S x^o L_S y^o$, and so $w^o L_S y^o$. (L_S is transitive.)

And R_3 *is irreflexive:* Since $w^o \sim w^o$, if $w^o R_3 w^o$, then $w^o L_S w^o$, which is impossible, as L_S is irreflexive.

Let $\mathcal{M}_3 = \langle W_2, R_3, P_2\rangle$.

We now need three lemmas.

Lemma 2

For every subset S of W_2 there is a truth-functional combination C of subsentences of A such that for every u in W_1, $\mathcal{M}_1 \vDash_u C$ iff $u^o \in S$.

Proof. If $S = W_2$, we take $C = \top$; if $S = \varnothing$, we take $C = \bot$. Otherwise, let $x_1^o, \ldots, x_p^o$ be all the elements of S and $y_1^o, \ldots, y_q^o$ be all the elements of $W_2 - S$.

For every pair $\langle u^o, v^o\rangle$ of *distinct* elements of W_2, choose a subsentence $D_{u^o v^o}$ of A that belongs to *exactly one* of u and v. (Since $u^o \neq v^o$ such a subsentence will always exist.) And for each i, j, $1 \leqslant i \leqslant p$, $1 \leqslant j \leqslant q$, let $C_{ij} = D_{x_i^o y_j^o}$ if $D_{x_i^o y_j^o} \in x_i$, and $= -D_{x_i^o y_j^o}$ if $D_{x_i^o y_j^o} \in y_j$. Then $\mathcal{M}_G \vDash_{x_i} C_{ij}$ and $\mathcal{M}_G \nvDash_{y_j} C_{ij}$. So $\mathcal{M}_1 \vDash_{x_i} C_{ij}$ and $\mathcal{M}_1 \nvDash_{y_j} C_{ij}$.

Let $C = (C_{11} \mathbin{\&} \cdots \mathbin{\&} C_{1q}) \vee \cdots \vee (C_{p1} \mathbin{\&} \cdots$

& C_{pq}). C is a truth-functional combination of subsentences of A. Moreover, $\mathcal{M}_1 \models_{x_i} C$ and $\mathcal{M}_1 \not\models_{y_j} C$.

Now if $u^o \in S$, then $u^o = x_i^o$, for some i. Since $\mathcal{M}_1 \models_{x_i} C$, by Lemma 1 $\mathcal{M}_2 \models_{u^o} C$, and by Lemma 1 again $\mathcal{M}_1 \models_u C$. And if $u^o \in W_2 - S$, then $u^o = y_j^o$ for some j. Since $\mathcal{M}_1 \not\models_{y_j} C$, $\mathcal{M}_2 \not\models_{u^o} C$, and so $\mathcal{M}_1 \not\models_u C$. Thus C has the property we want. ⊣

Lemma 3

If S is a nonempty subset of W_2, then for some a in W_1, $a^o \in S$ and for every x such that aR_1x, $x^o \notin S$.

Proof. Let S be a nonempty subset of W_2. By Lemma 2 there is a C such that $\mathcal{M}_1 \models_u C$ iff $u^o \in C$ (all u in W_1). We want to find an a in W_1 such that $\mathcal{M}_1 \models_a C$ and $\mathcal{M}_1 \models_a \Box -C$, for then $a^o \in S$, and, for all x such that aR_1x, $\mathcal{M}_1 \not\models_x C$, and so for all x such that aR_1x, $x^o \notin S$. But since S is nonempty, $w^o \in S$ for some w in W_1, and therefore $\mathcal{M}_1 \models_w C$. If $\mathcal{M}_1 \models_w \Box -C$, we are done, but if $\mathcal{M}_1 \not\models_w \Box -C$, then $\mathcal{M}_1 \not\models_w \Box(\Box -C \rightarrow -C)$ (this is the only place in the entire proof at which we appeal to the special axiom for G), and so for some a in W_1 such that wR_1a, $\mathcal{M}_1 \not\models_a \Box -C \rightarrow -C$, and so $\mathcal{M}_1 \models_a \Box -C$ and $\mathcal{M}_1 \models_a C$. ⊣

Lemma 4

For every subsentence B of A and every w^o in W_2, $\mathcal{M}_2 \models_{w^o} B$ if and only if $\mathcal{M}_3 \models_{w^o} B$.

Proof. Induction on the complexity of B. The only nontrivial case is that in which $B = \Box D$. Suppose that $\mathcal{M}_2 \models_{w^o} \Box D$. If $w^oR_3x^o$, then $w^oR_2x^o$, and so $\mathcal{M}_2 \models_{x^o} D$, whence, by the induction hypothesis, $\mathcal{M}_3 \models_{w^o} D$. We thus have that $\mathcal{M}_3 \models_{w^o} \Box D$.

For the converse, suppose that $\mathcal{M}_2 \not\models_{w^o} \Box D$. Then $\mathcal{M}_1 \models_w \Box D$ (Lemma 1), and so $\Box D \notin w$. Let $S = \{y^o | w^o \sim y^o\}$. Since $w^o \sim w^o$, S is a nonempty subset of W_2. By Lemma 3, for some a in W_1, $w^o \sim a^o$ and for every x

such that aR_1x, $w^o \not\sim x^o$. Since $w^o \sim a^o$, we have that $\Box D \notin a$, $\mathcal{M}_1 \nvDash_a \Box D$, and so for some b, aR_1b and $\mathcal{M}_1 \nvDash_b D$. So $\mathcal{M}_2 \nvDash_{b^o} D$ (Lemma 1), and then by the induction hypothesis, $\mathcal{M}_3 \nvDash_{b^o} D$. Since aR_1b, both $a^oR_2b^o$ and $w^o \not\sim b^o$. Since $w^o \sim a^o$, by the transitivity of R_2, $w^oR_2b^o$, and then since $w^o \not\sim b^o$, it follows that $w^oR_3b^o$. And since $\mathcal{M}_3 \nvDash_{b^o} D$ and $w^oR_3b^o$, we have that $\mathcal{M}_3 \nvDash_{w^o} \Box D$. ⊣

Since $\mathcal{M}_1 \nvDash_t A$, we have from Lemmas 1 and 4 that $\mathcal{M}_3 \nvDash_{t^o} A$. $\mathcal{M}_3$ is a model with a finite domain and a transitive and irreflexive accessibility relation. The completeness of G with respect to finite strict partial orderings has therefore been established: A modal sentence is a theorem of G if it is valid in all finite strict partial orderings. And since, as we saw, every finite strict partial ordering is transitive and well-capped, and 'if' in the last statement can be strengthened to 'if and only if'.

The completeness theorem for G implies that G has a noteworthy property: If $\vdash_G (\Box A_1 \vee \cdots \vee \Box A_n)$, then $\vdash_G A_m$ for some m, $1 \leqslant m \leqslant n$. For suppose that for each m, $\nvdash_G A_m$. Then by the completeness theorem, for each m there exists a transitive, well-capped model $\langle W_m, R_m, P_m \rangle$ such that for some w_m in W_m, $\langle W_m, R_m, P_m \rangle \nvDash_{w_m} A_m$. We may suppose that W_m is disjoint from W_k if $m \neq k$. (Standard device: replace each W_m by $\{\langle w, m \rangle | w \in W_m\}$ and redefine R_m and P_m accordingly.) Let a be some object that is in no W_m. Let $W = \{a\} \cup W_1 \cup \cdots \cup W_n$. Let $R = R_1 \cup \cdots \cup R_n \cup \{\langle a, w \rangle | w \in W_1 \cup \cdots \cup W_n\}$. And let P be such that $P(w,p) = P_m(w,p)$ whenever $w \in W_m$ (for all sentence letters p). Since the W_ms are all disjoint, R is transitive. And if $x_0Rx_1Rx_2Rx_3 \cdots$, then for some m, $x_1R_mx_2R_mx_3 \cdots$, which is impossible since R_m is well-capped; R itself is therefore well-capped. Theorem 1 of Chapter 5 implies that $\langle W, R, P \rangle \nvDash_{w_m} A_m$. Since aRw_m, it follows that $\langle W, R, P \rangle \nvDash_a \Box A_m$, for every m, $1 \leqslant m \leqslant n$. We therefore have

$$\langle W, R, P \rangle \nvDash_a (\Box A_1 \vee \cdots \vee \Box A_n),$$

whence $\nvdash_G (\Box A_1 \vee \cdots \vee \Box A_n)$.

Taking $n = 1$, we have that if $\vdash_G \Box A$, then $\vdash_G A$.

An application to Peano Arithmetic: Suppose that A_1 and A_2 are two modal sentences such that for every realization ϕ, $\vdash_{PA}$ Bew$[A_1{}^\phi]$ v Bew$[A_2{}^\phi]$. Then, of course, for every ϕ, Bew$[A_1{}^\phi]$ v Bew$[A_2{}^\phi]$ is true, and therefore for every ϕ, either $\vdash_{PA} A_1{}^\phi$ or $\vdash_{PA} A_2{}^\phi$. But something stronger holds. By Solovay's completeness theorem (cf. Chapter 12), $\vdash_G \Box A_1 \vee \Box A_2$, and therefore, as we have just seen, either $\vdash_G A_1$ or $\vdash_G A_2$. It thus follows that either for every ϕ, $\vdash_{PA} A_1{}^\phi$ or for every ϕ, $\vdash_{PA} A_2{}^\phi$. (We cannot prove that if for every ϕ, $\vdash_{PA} A_1{}^\phi$ v $A_2{}^\phi$, then either for every ϕ, $\vdash_{PA} A_1{}^\phi$ or, for every ϕ, $\vdash_{PA} A_2{}^\phi$: Let $A_1 = p$, $A_2 = -p$.)

Exercise

Show that we can improve our completeness theorem so that it reads: A is a theorem of G if A is valid in all finite, irreflexive trees; $\langle W, R\rangle$ is a finite, irreflexive tree if R is a transitive, irreflexive relation on a finite set W, $W = \{t\} \cup \{w|tRw\}$ for some t, and for all x, y, z, if xRz and yRz, then either xRy or $x = y$ or yRx. (A proof of this result is given in the next chapter.) [*Hint:* The last condition in the definition ensures that states of affairs of the sort:

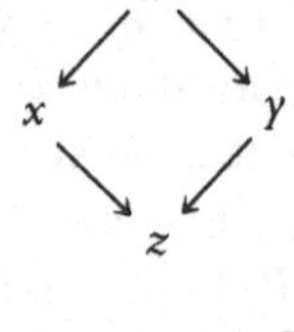

never occur. If A is not a theorem of G, then for some t, W, R, P, with W finite and R transitive and irreflexive, $\langle W,R,P\rangle \nvDash_t A$. Let $W' = \{t\} \cup \{w|tRw\}$, and let R' and P' be the restrictions of R and P to W'. Then $\langle W',R',P'\rangle \nvDash_t A$. Unsnarl $\langle W',R'\rangle$: Inductively replace occurrences of

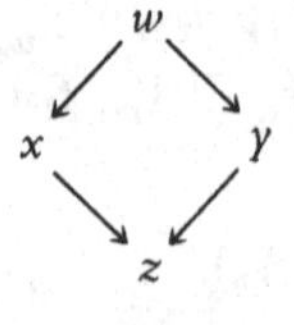

by occurrences of

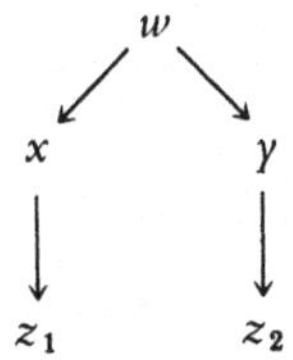

obtaining $\langle W'', R'' \rangle$ thereby. Define a suitable P'' along the way, by setting $P''(z_1,p) = P''(z_2,p) = P'(z,p)$.]

8

Trees for G

The method of truth-trees, due to Smullyan, is a proof procedure for propositional and predicate logic that is an attractive simplification of the proof procedure due to Beth called the method of semantic tableaux, which is in turn an adaptation of proof procedures due to Gentzen and Herbrand. Kripke showed how the method of semantic tableaux for the propositional calculus could be extended to provide completeness proofs for several systems of modal propositional logic. In the present chapter we shall adapt Kripke's methods to show how the method of trees may be extended to prove the completeness of G with respect to models $\langle W, R, P \rangle$, in which W is finite and R is a transitive and irreflexive relation on W.[1] The extension to G of the method of trees also supplies us with a quite practical decision procedure for G. We shall assume that the reader is already familiar with some presentation of the method of truth-trees for the propositional calculus, such as the one in Smullyan's *First-Order Logic* or Jeffrey's *Formal Logic: Its Scope and Limits*.

Let us first take a look at a few examples to see what our extension of the method of trees looks like.

We test a sentence for theoremhood in G by testing its negation for consistency with G. Let us test $\Box(\Box p \rightarrow p) \rightarrow \Box p$ for theoremhood (Example 1). In step

$$\vdash_G \Box(\Box p \to p) \to \Box p \quad ?$$

(1) $\sqrt{} -(\Box(\Box \to p) \to \Box p)$

(2) $\quad \sqrt{}\Box(\Box p \to p)$

$\quad\quad \sqrt{} -\Box p$

(3) $\quad -p$

$\quad\quad \Box p$

(4) $\sqrt{}\Box p \to p$

$\quad \Box(\Box p \to p)$

$-\Box p \diagup \quad \diagdown p$

(5)

$\times \quad\quad\quad \times$

$\times$

(6)

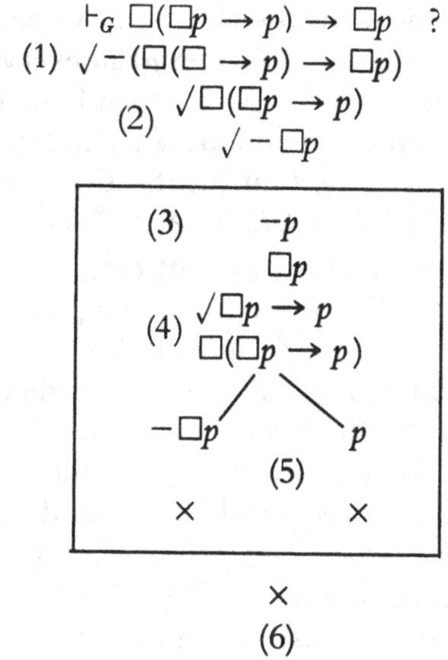

Example 1

(1) we write down its negation, $-(\Box(\Box p \to p) \to \Box p)$. In step (2) we apply the propositional calculus rules as many times as we can, inferring $\Box(\Box p \to p)$ and $-\Box p$ from $-(\Box(\Box p \to p) \to \Box p)$, and checking $-(\Box(\Box p \to p) \to \Box p)$ to indicate that we have finished with it. In step (3) we "open a window onto a possible world." Since $\vdash_G -\Box p \to \Diamond(-p \,\&\, \Box p)$ [cf. Theorem 10(b) of Chapter 1], we (guess how much space we will later need and) write

$-p$

$\Box p$

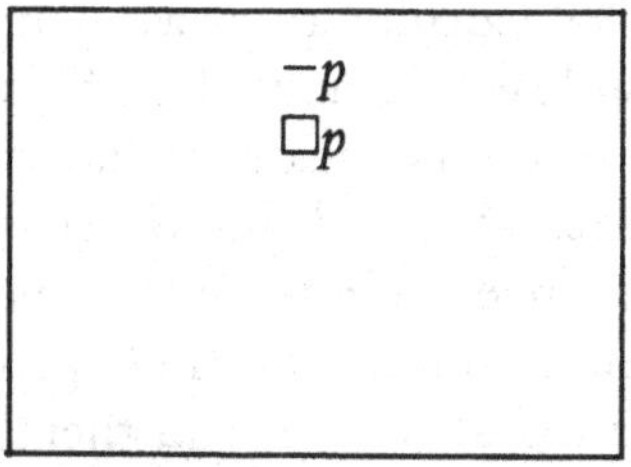

meaning: $\Diamond(-p \,\&\, \Box p)$. We then check $-\Box p$. Since there are no unchecked sentences of the form $-\Box A$, we have opened as many windows as possible, and we pass to step (4). We write $\Box p \rightarrow p$ and $\Box(\Box p \rightarrow p)$ in every window – there is only one – that is on an open branch on which $\Box(\Box p \rightarrow p)$ appears. When done we check the (outer) occurrence of $\Box(\Box p \rightarrow p)$. Our justification for doing so is that $\vdash_G \Box A \rightarrow (\Diamond B \leftrightarrow \Diamond(B \mathbin{\&} A \mathbin{\&} \Box A))$ [Theorem 10(e) of Chapter 1. In the present case, $A = \Box p \rightarrow p$ and $B = (-p \mathbin{\&} \Box p)$.] In step (5) we apply the propositional calculus rules inside the window as many times as possible, obtaining a closed tree inside the window. In step (6) we close the branch on which the window lies because there is a closed tree inside the window. Our justification here is that if $\vdash_G -D$ and $\vdash_G C \rightarrow \Diamond D$, then $\vdash_G \Box -D$, $\vdash_G -\Diamond D$, and so $\vdash_G -C$. Since all branches are closed, the tree is closed, and $-(\Box(\Box p \rightarrow p) \rightarrow \Box p)$ is not consistent with G, that is, $\Box(\Box p \rightarrow p) \rightarrow \Box p$ is a theorem of G (as of course we knew).

In Example 2, we negate the sentence that we are testing for theoremhood and apply the propositional calculus rules as many times as we can. Two of the sentences we obtain are $-\Box p$ and $-\Box\Box p$, and so we open two windows on the sole branch so far obtained. At the top of one of them we write $-p$ and $\Box p$; at the top of the other, $-\Box p$ and $\Box\Box p$. Since $\Box(p \vee \Box p)$ lies on a branch on which both of these windows occur, we write $p \vee \Box p$ and $\Box(p \vee \Box p)$ inside both the windows. We then apply the propositional calculus rules as many times as possible. In the top window, we have then finished: One branch is closed, but the other is open. In the bottom window, one branch is closed, the other is open, but we have not finished, since $-\Box p$ is on the open branch. We therefore open a window on this open branch: We write $-p$ and $\Box p$ at its top, and then, since $\Box\Box p$ and $\Box(p \vee \Box p)$ lie on the branch on which the window lies, we write down $\Box p$ and $\Box\Box p$, and $p \vee \Box p$ and $\Box(p \vee \Box p)$ inside the window. We then apply the

$\vdash_G \Box(p \vee \Box p) \rightarrow (\Box p \vee \Box\Box p)$?

$$\surd -(\Box(p \vee \Box p) \rightarrow (\Box p \vee \Box\Box p))$$

$$\surd \Box(p \vee \Box p)$$

$$\surd -(\Box p \vee \Box\Box p)$$

$$\surd -\Box p$$

$$\surd -\Box\Box p$$

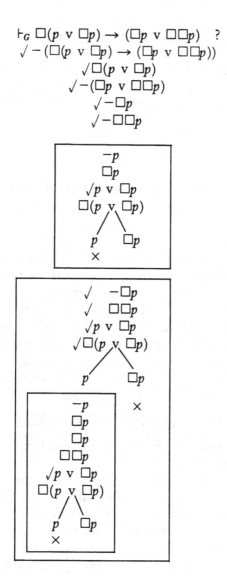

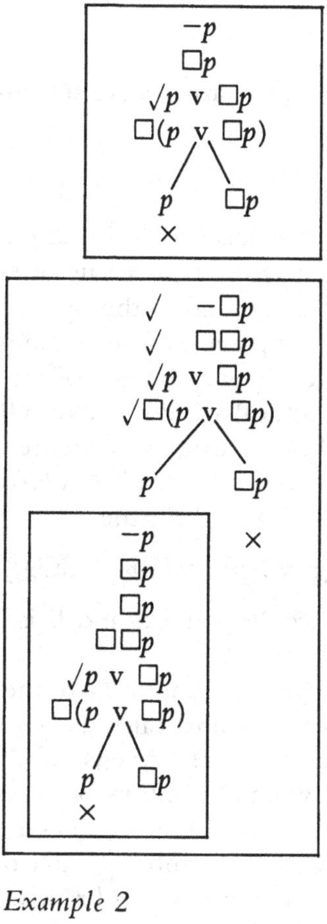

Example 2

propositional calculus rules again. One branch is closed, but the other remains open. There is then nothing more to do, and we have in fact constructed a model in which the sentence that we tested is invalid: There is a world of the model at which its negation is true. The model looks like this:

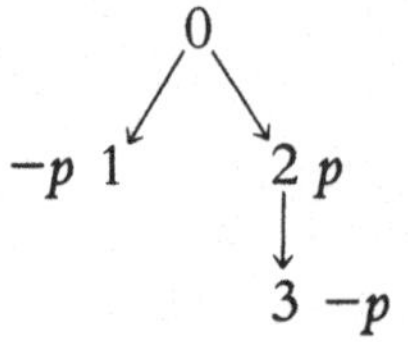

(The arrows are transitive.)

At world 3 of this model, p is false; at world 2, p is true; at world 1, p is false; and at world 0, p is true or false, at pleasure. Worlds 1 and 2 correspond to the open branches of the trees in the top and bottom windows, and world 3 corresponds to the open branch of the tree in the window within the (bottom) window. World 0 is the world at which the denial of the original sentence is true. So if we take $W = \{0,1,2,3\}$ and $R = \{\langle 0,1\rangle,\langle 0,2\rangle,\langle 2,3\rangle,\langle 0,3\rangle\}$, and let $P(1,p) = P(3,p) = \perp \neq P(2,p)$, then

$$\langle W,R,P\rangle \models_0 -(\Box(p \vee Lp) \rightarrow (\Box p \vee \Box\Box p)),$$

as an easy calculation shows. W is finite, and R is transitive and irreflexive.

We shall now describe the extension to G of the method of trees for the propositional calculus. We must first say what the (nesting-)*degree* of a tree is, what ‘closed’ and ‘open’ mean, what an *available* occurrence of a sentence is, and what the $-\Box$ and the $\Box$ rules are.

The *degree* of a tree is the least number greater than the degrees of all trees inside windows on branches of the tree. Thus a tree with no windows has degree 0, and a tree has degree $n + 1$ if and only if on at least one of its branches there is a window inside which there is a tree of degree n, and inside every window on a branch of the tree there is a tree of degree $\leqslant n$.

All trees in windows of a tree of degree n are of degree $<n$. So, inductively, we call a *tree* (of degree n) *closed* if all its branches are closed, and a *branch* (of a tree of degree n) *closed* if it either contains $\perp$ or contains some sentence and its negation or contains a window inside which there is a closed tree (which will be of degree $<n$). 'Open' (when said of a tree or a branch) means 'not closed'.

An occurrence of a sentence is *available* if it is unchecked and

(1) it lies on an open branch of a tree that is inside no window; or
(2) it lies on an open branch of a tree that is inside a window on an open branch of a tree that is inside *no* window; or
(3) it lies on an open branch of a tree that is inside a window on an open branch of a tree that is inside a window on an open branch of a tree that is inside *no* window; or . . . .

The* $-\Box$ *rule. If there is an available occurrence of a sentence $-\Box A$ on some open branch, write down a window containing the (one-branch) tree

$-A$
$\Box A$

on each open branch that contains the occurrence. Then check the occurrence.

The* $\Box$ *rule. If there are a window and an available occurrence of a sentence $\Box A$ on some open branch, write down both A and $\Box A$ at the bottom of every open branch of every tree inside every window on every open branch that contains the occurrence. Then check the occurrence.

Our procedure for developing trees for testing a sentence for theoremhood is shown in the flowchart

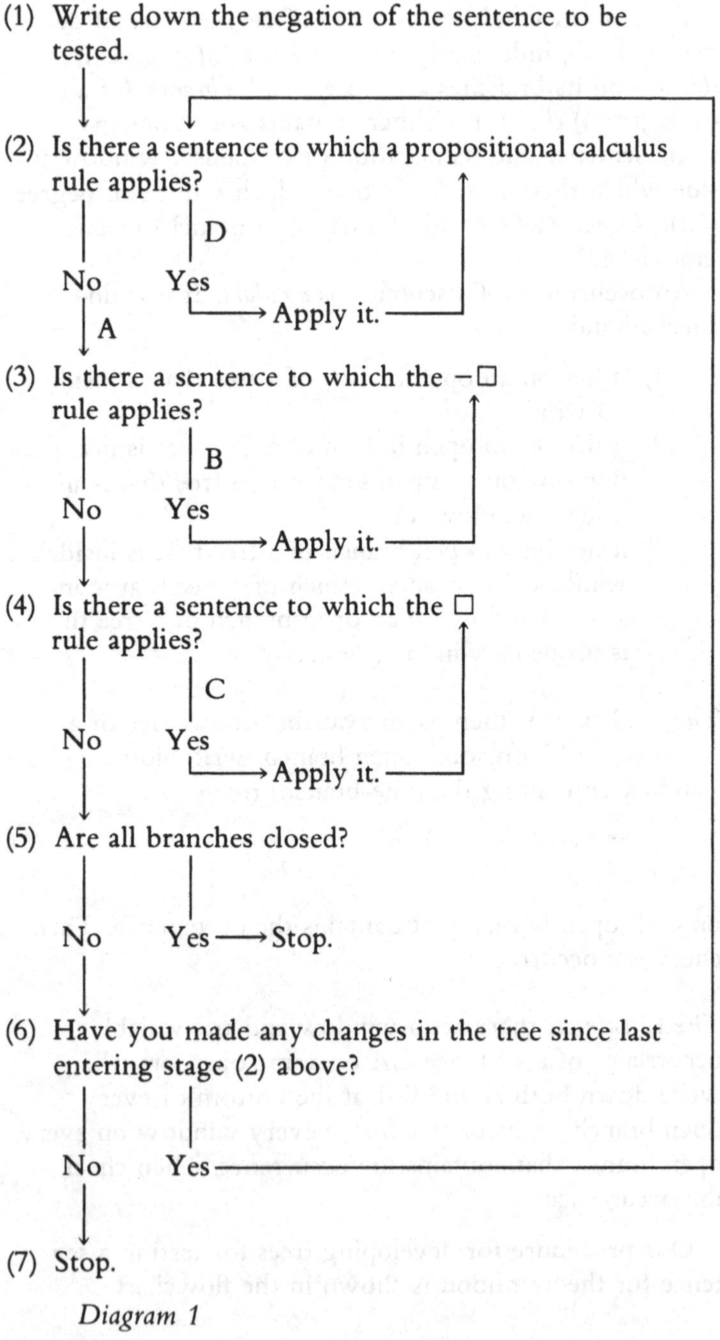
(1) Write down the negation of the sentence to be tested.
(2) Is there a sentence to which a propositional calculus rule applies?
D
No
Yes
Apply it.
A
(3) Is there a sentence to which the −□ rule applies?
B
No
Yes
Apply it.
(4) Is there a sentence to which the □ rule applies?
C
No
Yes
Apply it.
(5) Are all branches closed?
No
Yes
Stop.
(6) Have you made any changes in the tree since last entering stage (2) above?
No
Yes
(7) Stop.

Diagram 1

(Diagram 1) (cf. *Formal Logic*[2]). It consists in writing down the negation of the sentence, applying the propositional calculus rules as many times as possible to available occurrences of sentences, then applying the $-\Box$ rule as many times as possible, then applying the $\Box$ rule as many times as possible, then applying the propositional calculus rules as many times as possible, then applying the $-\Box$ rule, ... We thus develop a tree as far as we can by the propositional calculus rules, then open windows (and perhaps windows within windows, etc.), then add sentences to the one-branch trees inside those windows, then develop those trees as far as we can by the propositional calculus rules, then add windows to those trees, ...

We shall show that no matter which sentence we test by the procedure, we eventually stop at either (5) or (7), that if we stop at (5), then the sentence under test is a theorem, and that if we stop at (7), then there is a transitive and irreflexive model with a finite domain, in which the sentence under test is invalid. Since every theorem of G is valid in all such models, we shall then have shown that a sentence is a theorem of G if and only if it is valid in all transitive and irreflexive models with a finite domain. We shall also have shown that our procedure is a decision procedure for theoremhood in G.

We first show that the procedure always comes to a stop.

Let us observe that each time we traverse arrow B, we reduce the complexity or number of sentences to which the $-\Box$ rule applies; in following our procedure, we therefore never make an uninterrupted infinite series of traversals of arrow B. (Similarly, in following the procedure of Jeffrey, we never make an uninterrupted infinite series of applications of the rule for existential quantifiers.) We likewise never make an uninterrupted infinite series of traversals of arrow C, or of arrow D. It follows that if we apply our procedure to test a sentence and never stop, then we traverse arrow A infinitely many times. We shall show that our procedure always stops by showing that no matter which sentence H we use it to test, the number of times we traverse arrow A is at most *two* greater than the number of subsentences of H that are of the form $\Box J$.

Let us call a traversal of arrow B that immediately follows a traversal of arrow A an *AB-traversal*. Immediately after we have completed our first AB-traversal, we shall have opened a window on an open branch of a tree T_1 with $-H$ at its top; there is an open tree T_2 in the window. In general, immediately after we have completed our nth AB-traversal, we shall have obtained at least one sequence of open trees $T_1, \ldots, T_n, T_{n+1}$, with $-H$ at the top of T_1, such that for each i, $1 \leqslant i \leqslant n$, T_{i+1} is in a window on an open branch b_i of T_i.

But if we have obtained such a sequence $T_1, \ldots, T_n, T_{n+1}$, then there exists a sequence of subsentences $\Box J_1, \ldots, \Box J_n$ of H such that for each i, $1 \leqslant i \leqslant n$, all of $\Box J_1, \ldots, \Box J_{i-1}$ and $-\Box J_i$ occur on some open branch b_i of T_i (see Diagram 2). As each b_i is open, $\Box J_i$ is identical to none of $\Box J_1, \ldots, \Box J_{i-1}$. It follows that there are at least n subsentences $\Box J_1, \ldots, \Box J_n$ of H.

Contraposing, we see that if there are at most $n - 1$ subsentences of H of the form $\Box J$, then we shall not complete n AB-traversals.

Notice, moreover, that if we traverse arrow A without immediately thereafter traversing arrow B, then we never again traverse arrows C, D, or B, for there are no new windows in which to write down sentences in accordance with the $\Box$ rule, and the propositional calculus rules have been applied as many times as possible. So if we traverse A and do not thereupon traverse B, we traverse A at most once more, after having made the big loop from (6) back to (2).

We conclude that if there are $n - 1$ subsentences of H of the form $\Box J$, then we traverse arrow A at most $n + 1$ times. (For we can traverse B immediately after tra-

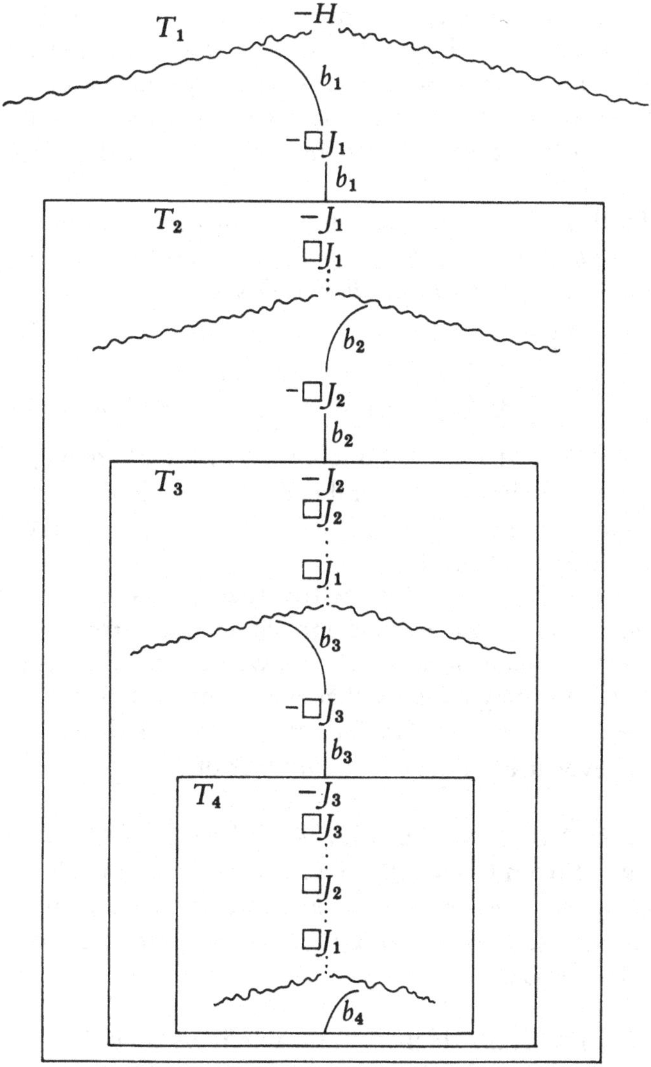

Diagram 2

versing A at most $n - 1$ times, and if we traverse A after those $n - 1$ times, then we can traverse A but once again.) Thus our procedure always comes to a stop.

We now want to show that if we apply our procedure to H and stop at (5), producing a closed tree, then H is a theorem of G.

To this end, we define (simultaneously) the notions of the *characteristic sentence* (T) *of a tree* T and the *characteristic sentence* (b) *of a branch* b as follows:

$$(T) = \bigvee\{(b) \mid b \text{ is a branch of } T\};$$
$$(b) = \bigwedge\{A \mid A \text{ is a sentence on } b\} \;\&\; \bigwedge\{\Diamond(T) \mid T \text{ is a tree in a window on } b\}.$$

(Here '$\bigvee$' means 'the disjunction of all members of'; '$\bigwedge$', 'the conjunction of all members of'.) The definition is not circular: Trees in windows on branches of T have lower degrees than T.

Suppose first that U is the tree that results from a tree T when one of our rules is applied *to an occurrence of a sentence A on a branch b of T.* We want to see that $\vdash_G (T) \rightarrow (U)$. We may suppose that A is either $-\Box B$, $\Box B$, $(B \rightarrow C)$, or $-(B \rightarrow C)$, for some sentences B, C. But whichever it is, A will be a conjunct of (b).

Case (*i*) *A is* $-\Box B$. Then if c is the branch of U that is obtained from b when the $-\Box$ rule is applied to the occurrence of A, then $(c) = (b) \;\&\; \Diamond(-B \;\&\; \Box B)$. But since $\vdash_G -\Box B \rightarrow \Diamond(-B \;\&\; \Box B)$, $\vdash_G (b) \rightarrow (c)$, and so $\vdash_G (T) \rightarrow (U)$.

Case (*ii*) *A is* $\Box B$. If there are n windows on b, let $D_1, \ldots, D_n$ be the characteristic sentences of the trees in these windows. Then $\Diamond D_1, \ldots, \Diamond D_n$ will be among the conjuncts of b. If c is the branch of U that is obtained from b when the $\Box$ rule is applied to the occurrence of A, then (c) is obtained from (b) by replacing each $\Diamond D_i$ by $\Diamond(D_i \;\&\; B \;\&\; \Box B)$. But since $\vdash_G \Box B \;\&\; \Diamond D_i$

$\rightarrow \Box B$ & $\Diamond(D_i$ & B & $\Box B)$ [cf. Theorem 10(e) of Chapter 1], we again have that $\vdash_G (b) \rightarrow (c)$, and so $\vdash_G (T) \rightarrow (U)$.

Case (iii) A is $(B \rightarrow C)$. After the $\rightarrow$ rule has been applied to the occurrence of A, b will have split into two branches of U, d and e, with $(d) = (b)$ & $-B$ and $(e) = (b)$ & C. Since $\vdash_G (b) \rightarrow (d) \vee (e)$, we again have that $\vdash_G (T) \rightarrow (U)$.

Case (iv) A is $-(B \rightarrow C)$. If c is obtained from b when the $- \rightarrow$ rule is applied to the occurrence of A, then $(c) = (b)$ & B & $-C$. Again, $\vdash_G (b) \rightarrow (c)$, and $\vdash_G (T) \rightarrow (U)$.

In all cases, then, $\vdash_G (T) \rightarrow (U)$. If T is in a window on a branch e of a tree X, and U, f, and Y are the tree, branch, and tree that result from T, e, and X when a rule is applied to an occurrence of a sentence in T, then, as we have seen, $\vdash_G (T) \rightarrow (U)$, and so $\vdash_G \Diamond(T) \rightarrow \Diamond(U)$, $\vdash_G (e) \rightarrow (f)$, and therefore $\vdash_G (X) \rightarrow (Y)$.

By induction (on degree) we infer that if U is the tree that results from a tree T when one of the rules is applied *anywhere*, then $\vdash_G (T) \rightarrow (U)$. Furthermore, if V is a tree that contains just one branch, and that branch contains just the one sentence $-H$, then $(V) = -H$. It follows by induction (on the number of traversals of arrows) that if T is the tree produced when we stop, then $\vdash_G -H \rightarrow (T)$. And since $\vdash_G -\Diamond A$ if $\vdash_G -A$, it follows by induction (on degree) that if T is a closed tree, then $\vdash_G -(T)$: For if b is a branch of T, then either $\perp$ is on b, or some sentence and its negation are, or some window on b contains a closed tree of lower degree, and in each of these cases, $\vdash_G -(b)$.

Thus if T is the tree generated by our procedure when it stops and T is closed, then, since $\vdash_G -H \rightarrow (T)$ and $\vdash_G -(T)$, $\vdash_G H$.

It remains to be shown that if we apply our procedure to H and stop at (7) (see Diagram 1), producing an open tree T, then there is a model $\langle W, R, P\rangle$, with W finite and R transitive and irreflexive, in which H is invalid. W will be a certain set of open branches obtained from T.

If U is an open tree, then U contains at least one open branch. If b is an open branch, then every tree in every window on b is open. Let us abbreviate 'the leftmost open branch of U' by '$L(U)$'. We take W to be the smallest set $\mathcal{D}$ that contains $L(T)$ and contains $L(U)$ whenever U is a tree in a window on a member of $\mathcal{D}$.

The degree of T is finite, and so W is a finite set.

We define a binary relation S by: bSc iff $b,c \in W$ and c is a branch of a tree in a window on b.

Let R be the ancestral S_* of S. (bRc iff for some $a_1, \ldots, a_n$, $b = a_1 S \cdots S a_n = c$.) R is transitive, as the ancestral of a relation is always a transitive relation. R is certainly irreflexive: No branch b of any tree in any window on any branch of any tree . . . is in a window on b.

Let P be the evaluator on W such that for all b in W and all sentence letters p, $P(b,p) = \top$ if and only if p lies on b.

Lemma
For every sentence A and every b in W, if A lies on b, then $\langle W,R,P\rangle \models_b A$, and if $-A$ lies on b, then $\langle W,R,P\rangle \not\models_b A$.

Proof. Induction on the complexity of A. There are four cases.

(*i*) $A = \bot$. $\bot$ does not lie on b since b is open. If $-\bot$ lies on b, then $\langle W,R,P\rangle \not\models_b \bot$.

(*ii*) *A is some sentence letter p.* If p lies on b, then $P(b,p) = \top$, and so $\langle W,R,P\rangle \models_b p$. If $-p$ lies on b, then since b

is open, p does not lie on b, and so $P(b,p) = \bot$, whence $\langle W,R,P \rangle \nvDash_b p$.

(*iii*) $A = (B \rightarrow C)$. If $(B \rightarrow C)$ lies on b, then the $\rightarrow$ rule has been applied to all occurrences of $(B \rightarrow C)$ on b, and therefore either $-B$ lies on b or C lies on b. By the induction hypothesis, either $\langle W,R,P \rangle \nvDash_b B$ or $\langle W,R,P \rangle \vDash_b C$. In either case, $\langle W,R,P \rangle \vDash_b (B \rightarrow C)$. If $-(B \rightarrow C)$ lies on b, then the $-\rightarrow$ rule has been applied to all occurrences of $-(B \rightarrow C)$ on b, and therefore both B and $-C$ occur on b. By the induction hypothesis, both $\langle W,R,P \rangle \vDash_b B$ and $\langle W,R,P \rangle \nvDash_b C$, whence $\langle W,R,P \rangle \nvDash_b (B \rightarrow C)$.

(*iv*) $A = \Box B$. If $-\Box B$ lies on b, then there is a window on b inside which there is a tree U at the very top of which the sentence $-B$ occurs. Since b is open, U is open, and therefore there is at least one open branch of U. Let c be the leftmost open branch of U. Then $c \in W$, bSc, and therefore bRc; $-B$ and $\Box B$ lie on c. So by the induction hypothesis $\langle W,R,P \rangle \nvDash_c B$, and, as bRc, $\langle W,R,P \rangle \nvDash_b \Box B$. Finally, suppose that $\Box B$ lies on b. If we can show that B lies on c whenever bRc, we shall be done, for then by the induction hypothesis, $\langle W,R,P \rangle \vDash_c B$ for all c such that bRc, and then $\langle W,R,P \rangle \vDash_b \Box B$. But observe that if $c,d \in W$, cSd, and $\Box B$ lies on c, then both B and $\Box B$ lie on d, for B and $\Box B$ lie on every open branch of every tree in every window on c. Thus if $b = a_1 \, S \, a_2 \cdots S \, a_n = c$, and $\Box B$ lies on b, then both B and $\Box B$ lie on c; thus if bRc, B lies on c. ⊣

Since $-H$ lies on every branch of T, $-H$ lies on $L(T)$. By the lemma, then, $\langle W,R,P \rangle \nvDash_{L(T)} H$, and therefore $\langle W,R,P \rangle$ is a model with a finite domain and a transitive and irreflexive accessibility relation in which H is invalid.

We have thus shown that a modal sentence H is a theorem of G if and only if H is valid in all models appropriate to G, if and only if our procedure, applied to *H*, stops at (5), if and only if our procedure, applied to *H*, does not stop at (7). The system G is therefore sound and complete with respect to models appropriate to G, and is also decidable.

If we examine the definition of *R* in the last part of our proof, we can see that $W = \{L(T)\} \cup \{b|L(T)Rb\}$, and for all b, c, d, if bRd and cRd, then either bRc or $b = c$ or cRb. Therefore $\langle W, R\rangle$ is a finite irreflexive tree, and the theorems of G are then just the sentences valid in all finite irreflexive trees. Furthermore, if *H* is not a theorem of G and there are *n* subsentences of *H* of the form $\Box J$, then *H* is invalid in some finite irreflexive tree $\langle W, R\rangle$ such that $b_0Rb_1R \cdot\cdot\cdot Rb_{n+1}$ for no $b_0, b_1, \ldots, b_{n+1}$ in *W*.

Exercises

1 Use the procedure to determine which of these are theorems of *G*:
(a) $\Box(\Box p \vee \Box - p) \rightarrow (\Box p \vee \Box - p)$.
(b) $\Box(\Box(p \mathbin{\&} q) \rightarrow p) \rightarrow \Box(\Box q \rightarrow p)$.
(c) $\Box(p \leftrightarrow (\Box p \rightarrow q)) \rightarrow \Box(p \leftrightarrow (\Box q \rightarrow q))$.
(d) $\Box(p \leftrightarrow (\Box(p \vee \Box\bot) \rightarrow \Box(p \rightarrow \Box\bot))$
$\rightarrow \Box(p \leftrightarrow (\Box\Box\Box\bot \rightarrow \Box\Box\bot))$.
(e) $\Box p \mathbin{\&} \Diamond q \rightarrow \Diamond(p \mathbin{\&} \Box\bot)$.
(f) $\Box(p \rightarrow \Box(p \rightarrow q)) \rightarrow \Box(p \rightarrow \Box q)$.
(g) $-\Box\bot \mathbin{\&} \Box(p \leftrightarrow -\Box p) \rightarrow (-\Box p \mathbin{\&} -\Box - p)$.

2 Modify our completeness proof to prove the completeness of other modal systems with respect to appropriate sorts of models.

3 (a) Show that if $\nvdash_K A$, then *A* is invalid in some $\langle W, R, P\rangle$, with *W* finite and *R* well-capped.
(b) Let *L* be the system obtained by adjoining to *K* the rule of inference: From $\vdash \Box A \rightarrow A$, infer $\vdash A$. Show that if $\vdash_L A$, then *A* is valid in every well-capped model.
(c) Conclude that *K* is closed under the rule: From $\vdash \Box A \rightarrow A$, infer $\vdash A$.

9

Calculating the truth-values of fixed points

We have already seen, or shall shortly see, that $\vdash_G \Box(p \leftrightarrow -\Box p) \rightarrow \Box(p \leftrightarrow -\Box\bot)$, $\vdash_G \Box(p \leftrightarrow \Box p) \rightarrow \Box(p \leftrightarrow \top)$, $\vdash_G \Box(p \leftrightarrow \Box -p) \rightarrow \Box(p \leftrightarrow \Box\bot)$, and $\vdash_G \Box(p \leftrightarrow -\Box -p) \rightarrow \Box(p \leftrightarrow \bot)$. From these facts about G we can conclude that for any sentence S of arithmetic (it is provable in arithmetic that) if S is equivalent to the assertion that S is unprovable, then S is equivalent to the assertion that arithmetic is consistent; if S is equivalent to the assertion that S is provable, then S is equivalent to $\top$; if S is equivalent to the assertion that S is disprovable, then S is equivalent to the assertion that arithmetic is inconsistent; and if S is equivalent to the assertion that S is consistent, then S is equivalent to $\bot$.

S is a sentence that is equivalent to the assertion that S is unprovable iff S is a Gödel sentence, that is, a $-\Box p$ fixed point; to the assertion that S is provable iff S is a Henkin sentence, that is, a $\Box p$ fixed point; to the assertion that S is disprovable iff S is a Jeroslow sentence, that is, a $\Box -p$ fixed point; and to the assertion that S is consistent iff S is a Rogers sentence, that is, a $-\Box -p$ fixed point (cf. Chapter 4.) Gödel, Henkin, Jeroslow, and Rogers sentences are all Gödelian fixed points, and (the arithmetizations of) the assertion that arithmetic is consistent, $\top$, the assertion that arithmetic is inconsistent, and $\bot$ are all deictic sentences.

$-\Box p$, $\Box p$, $\Box -p$, and $-\Box -p$ are all sentences that are modalized in p and contain no sentence letters other than p. And $-\Box\bot$, $\top$, $\Box\bot$, and $\bot$ are truth-functional combinations of lies, that is, of sentences of the form $\Box^n\bot$, $n \geqslant 0$.

We have just seen (two proofs of) a completeness theorem for G: The theorems of G are precisely those sentences that are valid in all models $\langle W, R, P\rangle$ in which W is finite and R is transitive and irreflexive. We are now going to use the completeness theorem for G to show that a very simple effective procedure, when applied to a sentence A that is modalized in p and contains no letters but p, yields a truth-functional combination H_A of lies such that $\vdash_G \Box(p \leftrightarrow A) \rightarrow \Box(p \leftrightarrow H_A)$. It follows quite directly that there are an effective procedure which, when applied to an A fixed point (A modalized in p), yields a deictic sentence that is equivalent to each A fixed point (and is itself an A fixed point), an effective procedure for calculating the truth-value of any given Gödelian fixed point, an effective procedure for calculating the truth-value of any given deictic sentence, and an effective procedure for calculating the provability-value of any given Gödelian fixed point.

HOW TO FIND H_A FROM A

Since A, as we suppose, is modalized in p and contains no sentence letters other than p, A is a truth-functional combination of sentences of the form $\Box D$. Let n = the number of subsentences of A that are of the form $\Box D$.

We now assign to each natural number i and sentence B containing no letters but p a truth-value, $\theta(i,B)$. We may suppose, an inductive hypothesis, that $\theta(j,B)$ has been defined for all $j < i$ and all B. We define: $\theta(i, \Box D) = \top$ iff for all $j < i$, $\theta(j,D) = \top$; $\theta(i,\bot) = \bot$; $\theta(i,B \rightarrow C) = \top$ iff either $\theta(i,B) = \bot$ or $\theta(i,C) = \top$; and $\theta(i,p) = \theta(i,A)$.

There are no natural numbers $j < 0$, and so $\theta(0, \Box D) = \top$. Moreover, $\theta(i, -B) = \top$ iff $\theta(i, B) = \bot$ (and similarly for the other connectives of the propositional calculus); and $\theta(i, \Diamond B) = \top$ iff for some $j < i$, $\theta(j, B) = \top$.

Every sentence containing no sentence letters other than p is a sentence of the form $\Box D$, or a truth-functional combination of such sentences, or the sentence A itself – which is such a truth-functional combination – or p, or a truth-functional combination of p and sentences of the form $\Box D$. $\theta(i, B)$ is therefore defined for every i and sentence B containing no letters other than p.

It is clear that there is an effective procedure by which we can calculate the value of θ for any given i, B.

To find H_A, first calculate $\theta(n, A)$. If $\theta(n, A) = \top$, take H_A to be

$$\bigwedge\{(\Box^j\bot \vee -\Box^{j+1}\bot) | j < n \; \& \; \theta(j, A) = \bot\},$$

but if $\theta(n, A) = \bot$, take H_A to be

$$\bigvee\{-(\Box^j\bot \vee -\Box^{j+1}\bot) | j < n \; \& \; \theta(j, A) = \top\}.$$

Evidently, H_A is effectively calculable from A and is a truth-functional combination of lies.

Examples

Suppose that $A = \Box p \to \Box -p$. Then $n = 2$. The following calculation gives $H_A = \Box\bot \vee -\Box^2\bot$:

	$\Box p$	$\Box -p$	$\Box p \to \Box -p$	p	$-p$
0	$\top$	$\top$	$\top$	$\top$	$\bot$
1	$\top$	$\bot$	$\bot$	$\bot$	$\top$
2	$\bot$	$\bot$	$\top$	$\top$	$\bot$

On each line we have set the truth-value of a sentence of the form $\Box D$ equal to $\top$ if and only if the value of D was $\top$ on all earlier lines (with the result that on line 0 both subsentences of A of the form $\Box D$ received $\top$), calculated the truth-values of truth-functional compounds from those of their components by appeal to the usual rules of the propositional calculus, and set the truth-value of p equal to that of A. On line 2, A received $\top$, and since the only line above line 2 on which A received $\bot$ was line 1, we took H_A to be $\Box\bot \vee -\Box^2\bot$.

The calculations for $A = \Box-p$ and $A = -\Box-p$ are

	$\Box-p$	p	$-p$
0	$\top$	$\top$	$\bot$
1	$\bot$	$\bot$	$\top$

$\Box-p$	$-\Box-p$	p	$-p$
$\top$	$\bot$	$\bot$	$\top$
$\top$	$\bot$	$\bot$	$\top$

These calculations yield $H_A = -(\Box^0\bot \vee -\Box^1\bot)$ and H_A = the empty disjunction. The first of these is equivalent (even in the propositional calculus) to $\Box\bot$; the second, to $\bot$.

We shall now verify that for all A modalized in p and containing no sentence letters but p, $\vdash_G \Box(p \leftrightarrow A) \rightarrow \Box(p \leftrightarrow H_A)$.

THE VERIFICATION

Let $A_1, \ldots, A_m$ be the sentence p and all the subsentences of A; let $\Box D_1, \ldots, \Box D_n$ be the n subsentences of A of the form $\Box D$ (if $n = 0$, A is a truth-functional compound of $\bot$), and let $C_l = \Box D_l$ for $1 \leqslant l \leqslant n$. '$k$' ranges over $\{1, \ldots, m\}$; 'l', over $\{1, \ldots, n\}$.

Lemma 1
$\theta(n, A) = \theta(r, A)$ for all $r > n$.

Proof. $\theta(i + 1, C_l) = \theta(i + 2, C_l)$ provided that $\theta(i, D_l) = \theta(i + 1, D_l)$, for then $\theta(j, D_l) = \top$ for all $j < i + 1$

if and only if $\theta(j, D_l) = \top$ for all $j < i + 2$. And $\theta(i, A_k) = \theta(i + 1, A_k)$ for all k if and only if $\theta(i, C_l) = \theta(i + 1, C_l)$ for all l, because each A_k is either a C_l or a truth-functional combination of C_ls, or p [but $\theta(i, p) = \theta(i, A)$ and A is a truth-functional combination of C_ls], or a truth-functional combination of p and some C_ls. So for every i,

(*) If $\theta(i, C_l) = \theta(i + 1, C_l)$ for all l,
then $\theta(i, A_k) = \theta(i + 1, A_k)$ for all k,
and so $\theta(i, D_l) = \theta(i + 1, D_l)$ for all l,
whence $\theta(i + 1, C_l) = \theta(i + 2, C_l)$ for all l,
and then $\theta(i + 1, A_k) = \theta(i + 2, A_k)$ for all k, ...

And if $i < j$ and $\theta(i, C_l) = \bot$, then [for some $h < i$, $\theta(h, D_l) = \bot$, and since $h < j$] $\theta(j, C_l) = \bot$. It follows that

(**) for some i, $0 \leqslant i \leqslant n$, $\theta(i, C_l) = \theta(i + 1, C_l)$ for all l.

[For otherwise, for every i, $0 \leqslant i \leqslant n$, there is an l such that $\theta(i, C_l) \neq \theta(i + 1, C_l)$, and therefore such that $\theta(i, C_l) = \top \neq \theta(j, C_l)$ for all $j \geqslant i + 1$; and therefore there are at least $n + 1$ sentences C_l, which is absurd – there are only n.]

Thus (*) and (**) give the lemma. ⊣

Suppose now that $\mathcal{M} = \langle W, R, P \rangle$, where R is a transitive and irreflexive relation and W is finite. Then $\mathcal{M}$ is appropriate to G, and $\breve{R}$ is well-founded.

It follows from the well-foundedness of $\breve{R}$ alone that if we wish to prove that all members of W have a property P, we need only show that each member w of W has P provided that all x such that wRx have P: For if the set S of members of W lacking P is nonempty, then, since $\breve{R}$ is well-founded, S contains a member w' that bears R to no member of S; but then $w' \in W$, w' lacks P, no x' such that $w'Rx'$ is in S, every x' such that $w'Rx'$ is in W and thus has P, and then (by what we assume to have been shown) w' has P, which is a contradiction. Proofs in which one concludes that all members of W have P by establishing that each member w of W has P if everything to which w bears R has P are called *proofs by $\breve{R}$-induction.*

We now define a function rk_R from members of W to natural numbers by: $rk_R(w)$ = the least natural number $> rk_R(x)$ for all x such that wRx. Since W is finite, there are only finitely many x such that wRx, and therefore if rk_R is well defined on all x such that wRx, it is well defined on w, too. By $\breve{R}$-induction, rk_R is well defined on all w in W. (Where the context makes it safe to do so, we shall drop subscript 'R'.)

So $rk(w) = 0$ iff for no x, wRx; if $rk(w) = i > 0$, then, since R is transitive, it follows by $\breve{R}$-induction that for every $h < i$, there is an x such that wRx and $rk(x) = h$.

Example

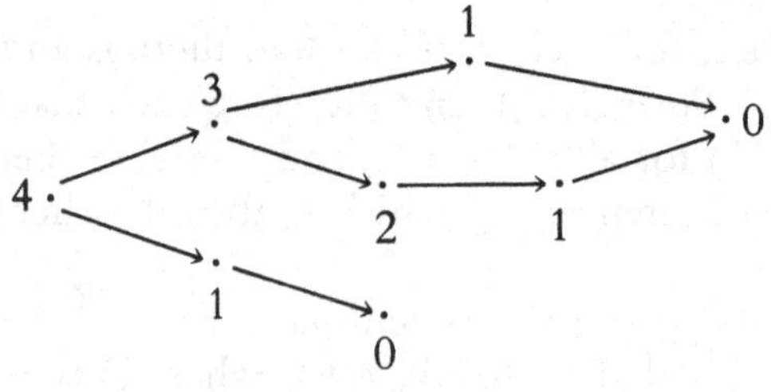

(The arrows are transitive.)

Moreover, since $rk(w) \geqslant j$ iff $\mathcal{M} \vDash_w \Diamond^j\top$, $rk(w) = j$ iff $\mathcal{M} \vDash_w (\Diamond^j\top \;\&\; -\Diamond^{j+1}\top)$, iff $\mathcal{M} \vDash_w -(\Box^j\bot \vee -\Box^{j+1}\bot)$.

Definition. $\mathcal{M}$, $= \langle W,R,P\rangle$, is *nice* iff for all w in W, $\mathcal{M} \vDash_w p \leftrightarrow A$.

Lemma 2

If $\mathcal{M}$ is nice and $rk(w) = i$, then for all k, $\mathcal{M} \vDash_w A_k$ iff $\theta(i, A_k) = \top$.

Proof. $\check{R}$-induction. Suppose the lemma holds for all x such that wRx. Then for all l, $\mathcal{M} \vDash_w C_l$ iff $\theta(i, C_l) = \top$: For if $\theta(i, C_l) = \bot$, then for some $h < i$, $\theta(h, D_l) = \bot$, and since $h < i$, for some x, wRx and $rk(x) = h$, whence by the inductive hypothesis, $\mathcal{M} \nvDash_x D_l$, and so $\mathcal{M} \nvDash_w C_l$; conversely, if $\mathcal{M} \nvDash_w C_l$, then for some x, wRx and $\mathcal{M} \nvDash_x D_l$, and so for some $h < i$, $rk(x) = h$, whence by the inductive hypothesis, $\theta(h, D_l) = \bot$, and so $\theta(i, C_l) = \bot$. Since A is a truth-functional combination of C_ls, $\mathcal{M} \vDash_w A$ iff $\theta(i, A) = \top$. $\mathcal{M} \vDash_w p$ iff $\mathcal{M} \vDash_w A$ since $\mathcal{M}$ is nice; $\theta(i,p) = \theta(i, A)$ by the definition of θ. Thus $\mathcal{M} \vDash_w p$ iff $\theta(i,p) = \top$. But every A_k is a truth-functional combination of p and some C_ls. ⊣

Lemma 3
Suppose that $\mathcal{M}$ is nice. Then $\mathcal{M} \vDash_w p$ iff $\mathcal{M} \vDash_w H_A$.

Proof. Let $i = rk(w)$. If $\theta(n, A) = \top$, then by Lemma 1, there are no $r > n$ such that $\theta(r, A) = \bot$, and so $\mathcal{M} \vDash_w H_A$ iff $\mathcal{M} \vDash_w (\Box^j\bot \vee -\Box^{j+1}\bot)$ for all j such that $\theta(j, A) = \bot$, iff $rk(w) \neq j$ for all j such that $\theta(j, A) = \bot$, iff $\theta(i, A) \neq \bot$, iff $\theta(i, A) = \top$. And if $\theta(n, A) = \bot$, then by Lemma 1 all j such that $\theta(j, A) = \top$ are $< n$, and so $\mathcal{M} \vDash_w H_A$ iff $\mathcal{M} \vDash_w -(\Box^j\bot \vee -\Box^{j+1}\bot)$ for some j such that $\theta(j, A) = \top$, iff $rk(w) = j$ for some j such that $\theta(j, A) = \top$, iff $\theta(i, A) = \top$. So $\theta(i, A) = \top$ iff $\mathcal{M} \vDash_w H_A$. By Lemma 2, $\mathcal{M} \vDash_w p$ iff $\theta(i,p) = \top$ (p is one of the A_ks); and by the definition of θ, $\theta(i,p) = \theta(i, A)$. ⊣

Definition. $\boxdot B = (\Box B \;\&\; B)$, for each modal sentence B.

If $\vdash_G B$, then $\vdash_G \boxdot B$. Since $\vdash_G \Box B \leftrightarrow (\Box\Box B \;\&\; \Box B)$, $\vdash_G \Box B \leftrightarrow \Box(\boxdot B)$. So if $\vdash_G \boxdot B \rightarrow C$, then by normality, $\vdash_G \Box(\boxdot B) \rightarrow \Box C$, and $\vdash_G \Box B \rightarrow \Box C$.

Theorem 1
$\vdash_G \boxdot(p \leftrightarrow A) \rightarrow (p \leftrightarrow H_A)$.

Proof. Suppose that $\mathcal{M} = \langle W,R,P\rangle$, W is finite, and R is transitive and irreflexive. Suppose further that $\mathcal{M} \models_w \boxdot(p \leftrightarrow A)$. Let $W' = \{w\} \cup \{x | wR_*x\}$; W' is finite. Since R is transitive, $R_* = R$, and $W' = \{w\} \cup \{x | wRx\}$. Define R' by: $yR'z$ iff y, $z \in W'$ and yRz. Since R is transitive and irreflexive, so is R'. Moreover, $W' = \{w\} \cup \{x | wR'x\}$. Define P' by $P'(y,q) = P(y,q)$ for all y in W' and all sentence letters q. Let $\mathcal{M}' = \langle W',R',P'\rangle$. By Theorem 1 of Chapter 5, $\mathcal{M}' \models_w \boxdot(p \leftrightarrow A)$. So $\mathcal{M}' \models_w p \leftrightarrow A$, and for all x such that $wR'x$, $\mathcal{M}' \models_x p \leftrightarrow A$. Thus for all y in W', $\mathcal{M}' \models_y p \leftrightarrow A$, that is, $\mathcal{M}'$ is nice. By Lemma 3, $\mathcal{M}' \models_w p \leftrightarrow H_A$, and by Theorem 1 of Chapter 5 again, $\mathcal{M} \models_w p \leftrightarrow H_A$. Thus if $\mathcal{M} \models_w \boxdot(p \leftrightarrow A)$, then $\mathcal{M} \models_w p \leftrightarrow H_A$. So $\mathcal{M} \models_w \boxdot(p \leftrightarrow A) \rightarrow (p \leftrightarrow H_A)$. But $\mathcal{M}$ was an arbitrary model with a finite domain and a transitive and irreflexive accessibility relation, and w was an arbitrary world of $\mathcal{M}$. Thus $\boxdot(p \leftrightarrow A) \rightarrow (p \leftrightarrow H_A)$ is valid in all such models, and by the completeness theorem for G, $\vdash_G \boxdot(p \leftrightarrow A) \rightarrow (p \leftrightarrow H_A)$.

Corollary 1
$\vdash_G \Box(p \leftrightarrow A) \rightarrow \Box(p \leftrightarrow H_A)$.

Proof. The corollary follows from Theorem 1 by the last remark that follows the definition of '$\boxdot B$'. ⊣

Suppose that A is modalized in p and contains no letters but p, and S is an A fixed point. Let ϕ be a realization such that $\Box(p \leftrightarrow A)^\phi$ is true. By Corollary 1, $\vdash_{PA} \Box(p \leftrightarrow A)^\phi \rightarrow \Box(p \leftrightarrow H_A)^\phi$, and thus if $\phi(p) = S$, then $\Box(p \leftrightarrow A)^\phi \rightarrow \mathrm{Bew}[(S \leftrightarrow H_a{}^\phi)]$ is true. Thus S is equivalent in arithmetic to the deictic sentence $H_A{}^\phi$. Thus there is an effective method for calculating, from any A fixed point (A modalized in p), a deictic sentence that is equivalent to each A fixed point.

Not only is $H_A{}^\phi$ equivalent to each A fixed point, it is itself an A fixed point. In fact, any sentence equivalent to an A fixed point is itself an A fixed point: If we write '$A(p)$' instead of 'A', by Theorem 5 of Chapter 1, we have $\vdash_G \Box(p \leftrightarrow q) \rightarrow \Box((p \leftrightarrow A(p)) \leftrightarrow (q \leftrightarrow A(q))$, and so by normality, $\vdash_G \Box(p \leftrightarrow q) \,\&\, \Box(p \leftrightarrow A(p)) \rightarrow \Box(q \leftrightarrow A(q))$. Thus if S is equivalent to S' and S is an $A(p)$ fixed point, then for some ϕ, $\phi(p) = S$ and $\Box(p \leftrightarrow A(p))^\phi$ is true. If we let ψ be such that $\psi(p) = S$ and $\psi(q) = S'$, then $\Box(p \leftrightarrow q)^\psi$ and $\Box(p \leftrightarrow A(p))^\psi$ are both true, and therefore $\Box(q \leftrightarrow A(q))^\psi$ is true. But then, if $\rho(p) = S'$, $\Box(p \leftrightarrow A(p))^\rho = \Box(q \leftrightarrow A(q))^\psi$, and so S' is an $A(p)$ fixed point.

To calculate the truth-value of the A fixed points (A modalized in p), we need only calculate $\theta(n, A)$, where n is the number of subsentences of A of the form $\Box D$. For if $\theta(n, A) = \top$, then at least one disjunct of every conjunct of H_A is the negation of a lie and $H_A{}^\phi$ is thus true; and, similarly, if $\theta(n, A) = \bot$, then $H_A{}^\phi$ is false.

There are two further corollaries to Theorem 1 that we should note.

Corollary 2
$\vdash_G \Box(p \leftrightarrow A) \rightarrow (\Box p \leftrightarrow \Box H_A)$.

Proof. Corollary 1 plus normality. ⊣

Corollary 3
If A is letterless, then $\vdash_G A \leftrightarrow H_A$.

Proof. If A is letterless, then p does not occur in A, and the result of substituting A for p in A is A. Therefore if we substitute A for p in Theorem 1, we obtain $\vdash_G \boxdot(A \leftrightarrow A) \rightarrow (A \leftrightarrow H_A)$. But of course $\vdash_G \boxdot(A \leftrightarrow A)$, and so $\vdash_G A \leftrightarrow H_A$. ⊣

Since H_A is a truth-functional combination of lies, the truth-value of any given deictic sentence can be effectively calculated, as can the provability value of any given A fixed point (A modalized in p). Of course, these results were established in Chapter 4 by proofs less complicated than the present one.

Historical note. Bernardi and Smoryński independently proved that for every A modalized in p and containing no letters other than p, there is a letterless H_A for which $\vdash_G \Box(p \leftrightarrow A) \rightarrow \Box(p \leftrightarrow H_A)$.[1] (For a proof of the correcteness of the algorithm described in this chapter that does not appeal to Lemma 1, see Smoryński, "Calculating self-referential statements.")

Exercises

1 $H_{(\Box(-p \rightarrow \Box\bot) \rightarrow \Box(p \rightarrow \Box\bot))} = ?$

2 (de Jongh). Let A, $= A(p)$, be modalized in p. Show that for every $\langle W, R, P\rangle$, with R transitive and well-capped, there exists a Q such that for all w in W, $\langle W, R, Q\rangle \models_w p \leftrightarrow A$ and for all $q \neq p$, $P(w,q) = Q(w,q)$. (*Hint:* Define Q by $\check{R}$-induction.) Then show that $\vdash_G H_A \leftrightarrow A(H_A)$. Then show that $\vdash_G \Box(p \leftrightarrow A) \leftrightarrow \Box(p \leftrightarrow H_A)$.

3 Define Con^n by Con^0 $= \top$; Con^{n+1} $=$ $-$Bew[$-$Con^n]. Let $\{p_i\}_{i<n}$ be a sequence of sentence letters. Prove that $\vdash_G \Box(\bigwedge_{i<n}(\Box p_i \rightarrow p_i) \rightarrow \Diamond^n\top) \rightarrow \Box(\Diamond^n\top \rightarrow \bigwedge_{i<n}(\Box p_i \rightarrow p_i))$ and deduce that every conjunction of n reflection principles that implies Con^n is equivalent to Con^n. (*Hint:* Assume that $\mathcal{M}$, $= \langle W, R, P\rangle$, is appropriate to G, $w \in W$, $\models_w \Box(\bigwedge_{i<n}(\Box p_i \rightarrow p_i) \rightarrow \Diamond^n\top)$, wRx, and $\models_x \Diamond^n\top$, but that $\not\models_x \bigwedge_{i<n}(\Box p_i \rightarrow p_i)$. Then for some $i < n$, $\not\models_x \Box p_i \rightarrow p_i$, and therefore $\models_x \Box p_i$. Since $\models_x \Diamond^n\top$, $rk(x) \geq n$, and so there is a sequence $\{x_j\}_{j<n}$ of members of W such that $x_k R x_j$ whenever $k < j < n$, $rk(x_j) = n - (j + 1)$, and xRx_j. Since wRx, wRx_j for all $j < n$, and so $\models_{x_j} \bigwedge_{i<n}(\Box p_i \rightarrow p_i) \rightarrow \Diamond^n\top$. Since $rk(x_j) < n$, $\not\models_{x_j} \Diamond^n\top$, and so $\not\models_{x_j} \bigwedge_{i<n}(\Box p_i \rightarrow p_i)$. Show by induction that for every $j < n$, there are at least $j + 1$ i such that $\models_{x_j} p_i$ & $\Box p_i$. Conclude that there are at least n i such that $\models_{x_{n-1}} p_i$ & $\Box p_i$, and hence that for all $i < n$ $\models_{x_{n-1}} \Box p_i \rightarrow p_i$, and obtain a contradiction.)

4 Show that no conjunction of fewer than n reflection principles implies Con^n.

10

Rosser's theorem

The first incompleteness theorem of Gödel asserts that if arithmetic is ω-consistent, then arithmetic is incomplete. But since every theorem of arithmetic is true, arithmetic, we may be certain, is consistent, and indeed ω-consistent, but not complete.

The second incompleteness theorem asserts that if arithmetic is consistent, then the assertion that arithmetic is consistent is not provable in arithmetic. Since arithmetic *is* consistent, the assertion that it is consistent is not provable in arithmetic. The assertion that arithmetic is inconsistent is also not provable in arithmetic, for it is false. So the assertion that arithmetic is consistent is undecidable in arithmetic. And since it is provable in arithmetic that if arithmetic is incomplete, then arithmetic is consistent, the assertion that arithmetic is incomplete is also not provable in arithmetic.

From what assumptions can we prove the incompleteness of arithmetic in arithmetic? Because the first incompleteness theorem can be proved in arithmetic, one correct, but weak, answer is, "From the assumption of ω-consistency." We know that we can prove in arithmetic that if consistency is undecidable, then arithmetic is incomplete. Can we show that the undecidability of consistency follows in arithmetic from the assumption of consistency and thereby improve our answer to "From the assumption of consistency"? One-half of what would be required has already been shown: Since the second incompleteness theorem can also be proved in arithmetic, we know that the unprovability of consistency follows from the assumption of consistency. Can we also show the other half, that the unprovability of inconsistency, also follows from the assumption of consistency?

We cannot: $-\text{Bew}[\bot] \to -\text{Bew}[\text{Bew}[\bot]]$ is not a theorem of arithmetic; for otherwise $\text{Bew}[\text{Bew}[\bot]] \to \text{Bew}[\bot]$ is a theorem, and therefore by Löb's theorem, $\text{Bew}[\bot]$ is also a theorem, which is absurd. Thus the undecidability of consistency is not deducible in arithmetic from the assumption of consistency.

The (arithmetization of the) consistency statement is a deictic sentence. According to Theorem 2 of Chapter 4, for no letterless modal sentence A, $\vdash_G -\Box\bot \to (-\Box A \mathbin{\&} -\Box - A)$. Since we know that for all letterless B, $\vdash_G B$ iff $\vdash_{PA} B^\phi$, we can conclude that for no *deictic* sentence S does the undecidability of S follow in arithmetic from the assumption of consistency. We cannot therefore expect to show that incompleteness follows from consistency by finding a deictic sentence whose undecidability can be shown to follow from consistency.

The consistency statement is a Gödelian fixed point; in fact, it is a $-\Box p$ fixed point. Gödel, of course, proved the first incompleteness theorem by finding a $-\Box p$ fixed point, showing it to be unprovable if arithmetic is consistent, and showing it to be irrefutable if arithmetic is ω-consistent. Since, as we saw in Chapter 9, the Gödelian fixed points are precisely the sentences equivalent to deictic sentences, we cannot hope to find a Gödelian fixed point whose undecidability follows from consistency, and *thereby* show that incompleteness follows from consistency.

If there is no deictic sentence whose undecidability follows from consistency, are there *any* sentences whose undecidability follows from consistency? Rosser showed that there are, and hence that incompleteness follows from consistency. In a moment we shall find such a sentence. But before we do, let us see that the results of Chapter 12 will put us in a position to answer this question simply by applying an algorithm that will be described there. In Chapter 12 we shall prove that for all modal sentences A, $\vdash_{G*} A$ if and only if for all interpretations ϕ, A^{ϕ} is true; we shall also show how to decide whether or not $\vdash_{G*} A$, for any given A. Now there is no sentence whose undecidability follows from consistency if and only if for all realizations ϕ, $-\Box(-\Box\bot \rightarrow (-\Box p \;\&\; -\Box -p))^{\phi}$ is true. So there is a sentence whose undecidability follows from consistency if and only if $\nvdash_{G*} -\Box(-\Box\bot \rightarrow (-\Box p \;\&\; -\Box -p))$, and by applying the algorithm for G^* we can tell whether or not this is the case.

How might we find a sentence whose undecidability follows from consistency, that is, a sentence R such that $\vdash_{\mathrm{PA}} -\mathrm{Bew}[\bot] \rightarrow (-\mathrm{Bew}[R] \;\&\; -\mathrm{Bew}[-R])$? Suppose[1] that it is possible to find a predicate $P(x)$ such that for all sentences S, it is provable in arithmetic from the assumption that arithmetic is consistent that

(1) if $\vdash_{\mathrm{PA}} S$, then $\vdash_{\mathrm{PA}} P(\ulcorner S \urcorner)$; and
(2) if $\vdash_{\mathrm{PA}} -S$, then $\vdash_{\mathrm{PA}} -P(\ulcorner S \urcorner)$.

Let R be a fixed point of $-P(x)$. Then it is provable in arithmetic that

(3) $\vdash_{\mathrm{PA}} R \leftrightarrow -P(\ulcorner R \urcorner)$.

It is thus provable from consistency that if $\vdash_{\mathrm{PA}} R$, then, by (1), $\vdash_{\mathrm{PA}} P(\ulcorner R \urcorner)$ and, by (3), $\vdash_{\mathrm{PA}} -P(\ulcorner R \urcorner)$; and it is also provable from consistency that if $\vdash_{\mathrm{PA}} -R$, then, by (2), $\vdash_{\mathrm{PA}} -P(\ulcorner R \urcorner)$ and, by (3), $\vdash_{\mathrm{PA}} P(\ulcorner R \urcorner)$. It is thus provable from consistency that if either R or $-R$ is provable then PA is inconsistent, and hence provable from consistency that neither R nor $-R$ is provable, that is, $\vdash_{\mathrm{PA}} -\mathrm{Bew}[\bot] \rightarrow (-\mathrm{Bew}[R] \;\&\; -\mathrm{Bew}[-R])$, as desired.

How might we find such a predicate $P(x)$? Recall the

formula $\mathrm{Pf}(y,x)$ and the primitive recursive term $\mathrm{Neg}(x_1)$. For every natural number n and sentence S of arithmetic, if n is the Gödel number of a proof of S, then $\vdash_{PA} \mathrm{Pf}(\mathbf{n},\ulcorner S\urcorner)$, but if not, then $\vdash_{PA} -\mathrm{Pf}(\mathbf{n},\ulcorner S\urcorner)$; and $\vdash_{PA} \mathrm{Neg}(\ulcorner S\urcorner) = \ulcorner -S\urcorner$. As arithmetic is consistent, there do not exist proofs of S and $-S$ for any sentence S. Consider the sequence of proofs in arithmetic, enumerated in the order of their Gödel numbers. Then if S is a theorem, then $-S$ has no proof, and so S has a proof that is earlier in the enumeration than every proof of $-S$. And if $-S$ is a theorem, then the first proof that is a proof either of S or of $-S$ is a proof of $-S$, and so it is not the case that S has a proof that is earlier than every proof of $-S$. We are thus led to consider the predicate $\mathrm{Prov}(x)$:

$$\exists y(\mathrm{Pf}(y,x) \ \&\ \forall z(z < y \rightarrow -\mathrm{Pf}(z,\mathrm{Neg}(x)))).$$

Let S be a sentence of arithmetic. Then the following argument, which can be formalized in PA, shows that it is provable from consistency that if $\vdash_{PA} S$, then $\vdash_{PA} \mathrm{Prov}(\ulcorner S\urcorner)$, and if $\vdash_{PA} -S$, then $\vdash_{PA} -\mathrm{Prov}(\ulcorner S\urcorner)$.

Suppose that arithmetic is consistent. If $\vdash_{PA} S$, then some number n is the Gödel number of a proof of S, and therefore $\vdash_{PA} \mathrm{Pf}(\mathbf{n},\ulcorner S\urcorner)$. By consistency, no number is the Gödel number of a proof of the negation of S, and therefore $\vdash_{PA} -\mathrm{Pf}(\mathbf{0},\mathrm{Neg}(\ulcorner S\urcorner))$, . . . , $\vdash_{PA} -\mathrm{Pf}(\mathbf{n-1}, \mathrm{Neg}(\ulcorner S\urcorner))$. Since

$$\vdash_{PA} \forall z(z < \mathbf{n} \rightarrow (z = \mathbf{0} \vee \cdots \vee z = \mathbf{n-1})),$$

we have

$$\vdash_{PA} \forall z(z < \mathbf{n} \rightarrow -\mathrm{Pf}(z,\mathrm{Neg}(\ulcorner S\urcorner))),$$ and so

$$\vdash_{PA} \mathrm{Pf}(\mathbf{n},\ulcorner S\urcorner) \ \&\ \forall z(z < \mathbf{n} \rightarrow -\mathrm{Pf}(z,\mathrm{Neg}(\ulcorner S\urcorner))),$$

whence

$$\vdash_{PA} \exists y(\mathrm{Pf}(y,\ulcorner S\urcorner) \ \&\ \forall z(z < y \rightarrow -\mathrm{Pf}(z,\mathrm{Neg}(\ulcorner S\urcorner)))),$$

that is, $\vdash_{PA} \mathrm{Prov}(\ulcorner S\urcorner)$.

If $\vdash_{PA} -S$, then some number n is the Gödel number of a proof of $-S$, and therefore $\vdash_{PA} \mathrm{Pf}(\mathbf{n}, \mathrm{Neg}(\ulcorner S \urcorner))$. By consistency, no number is the Gödel number of a proof of S, and therefore $\vdash_{PA} -\mathrm{Pf}(\mathbf{0}, \ulcorner S \urcorner), \ldots, \vdash_{PA} -\mathrm{Pf}(\mathbf{n}, \ulcorner S \urcorner)$. Since $\vdash_{PA} \forall y((y \neq \mathbf{0} \;\&\; \cdots \;\&\; y \neq \mathbf{n}) \rightarrow \mathbf{n} < y)$, we have $\vdash_{PA} \forall y(\mathrm{Pf}(y, \ulcorner S \urcorner) \rightarrow \mathbf{n} < y)$, and so

$$\vdash_{PA} \forall y(\mathrm{Pf}(y, \ulcorner S \urcorner) \rightarrow (\mathbf{n} < y \;\&\; \mathrm{Pf}(\mathbf{n}, \mathrm{Neg}(\ulcorner S \urcorner)))),$$

whence

$$\vdash_{PA} \forall y(\mathrm{Pf}(y, \ulcorner S \urcorner) \rightarrow \exists z(z < y \;\&\; \mathrm{Pf}(z, \mathrm{Neg}(\ulcorner S \urcorner)))),$$

and then by the predicate calculus, $\vdash_{PA} -\mathrm{Prov}(\ulcorner S \urcorner)$.

We have thus found that the predicate Prov(x) is a satisfactory $P(x)$, and therefore shown that Rosser's theorem, which asserts that if arithmetic is consistent, then arithmetic is incomplete, can also be proved in arithmetic.

$\vdash_{PA} \top$, and so by (1), $\vdash_{PA} \mathrm{Prov}(\ulcorner \top \urcorner)$, and consequently, $\vdash_{PA} \top \leftrightarrow \mathrm{Prov}(\ulcorner \top \urcorner)$. We also have, however, that $\vdash_{PA} -\bot$, and so by (2), $\vdash_{PA} -\mathrm{Prov}(\ulcorner \bot \urcorner)$ [cf. $\nvdash_{PA} -\mathrm{Bew}(\ulcorner \bot \urcorner)$] and consequently, $\vdash_{PA} \bot \leftrightarrow \mathrm{Prov}(\ulcorner \bot \urcorner)$. Prov($x$) is thus a predicate of which $\top$ and $\bot$, which are inequivalent, are fixed points. [By contrast, the fixed points of Bew(x) are precisely the sentences equivalent to $\top$.]

By the predicate calculus, $\vdash_{PA} \forall x(\mathrm{Prov}(x) \rightarrow \mathrm{Bew}(x))$. And by utilizing the fact that it is provable in arithmetic that if arithmetic is consistent, then no proof of any sentence has a greater Gödel number than some proof of the negation of the sentence, one can easily show that $\vdash_{PA} -\mathrm{Bew}[\bot] \rightarrow \forall x(\mathrm{Bew}(x) \leftrightarrow \mathrm{Prov}(x))$. In fact, since $\vdash_{PA} -\mathrm{Prov}(\ulcorner \bot \urcorner)$, $\vdash_{PA} -\mathrm{Bew}[\bot] \leftrightarrow \forall x(\mathrm{Bew}(x) \leftrightarrow \mathrm{Prov}(x))$. Thus Bew($x$) and Prov($x$) are coextensive predicates, but their coextensiveness is not provable in PA. And although Prov(x) is coextensive with Bew(x), $-\mathrm{Prov}(\ulcorner \bot \urcorner)$ does not express the consistency of PA; at best, it merely asserts that every proof of $\bot$ that there might be occurs later than some proof of $-\bot$. [Another example of a predicate $D(x)$ coextensive with Bew(x)

and such that $\vdash_{PA} -D(\ulcorner\bot\urcorner)$ is the predicate (Bew(x) & $x \neq \ulcorner\bot\urcorner$).]

Fixed points of $-$Prov(x) are called *Rosser sentences*. We have seen that if R is a Rosser sentence, then $\vdash_{PA} -\text{Bew}[\bot] \rightarrow (-\text{Bew}[R] \;\&\; -\text{Bew}[-R])$. It follows that no deictic sentence is equivalent to a Rosser sentence. Theorem 3 of Chapter 4 tells us that there is no non-provable deictic sentence that is strictly weaker than $-$Bew[$\bot$]. We shall now see that every Rosser sentence is strictly weaker than $-$Bew[$\bot$].

Let R be a Rosser sentence. Then $\vdash_{PA} R \leftrightarrow -\text{Prov}(\ulcorner R\urcorner)$, and so $\vdash_{PA} -\text{Prov}(\ulcorner R\urcorner) \rightarrow R$. But $\vdash_{PA} \forall x(\text{Prov}(x) \rightarrow \text{Bew}(x))$, and then, since $\vdash_{PA} -\text{Bew}[\bot] \rightarrow -\text{Bew}[R]$, we have $\vdash_{PA} -\text{Bew}[\bot] \rightarrow R$. If, conversely, $\vdash_{PA} R \rightarrow -\text{Bew}[\bot]$, then $\vdash_{PA} \text{Bew}[\bot] \rightarrow -R$, and so $\vdash_{PA} \text{Bew}[\text{Bew}[\bot]] \rightarrow \text{Bew}[-R]$, whence $\vdash_{PA} -\text{Bew}[-R] \rightarrow -\text{Bew}[\text{Bew}[\bot]]$; but since $\vdash_{PA} -\text{Bew}[\bot] \rightarrow -\text{Bew}[-R]$, we have $\vdash_{PA} -\text{Bew}[\bot] \rightarrow -\text{Bew}[\text{Bew}[\bot]]$, and so $\vdash_{PA} \text{Bew}[\text{Bew}[\bot]] \rightarrow \text{Bew}[\bot]$, whence by Löb's theorem $\vdash_{PA} \text{Bew}[\bot]$, which is absurd. Thus each Rosser sentence follows from consistency but not vice versa.

(What about $-R$ and $-$Bew[$\bot$]? Since $\vdash_{PA} -\text{Bew}[\bot] \rightarrow R$, if $\vdash_{PA} -R \rightarrow -\text{Bew}[\bot]$, then $\vdash_{PA} -R \rightarrow R$, $\vdash_{PA} R$, and PA is inconsistent; and if $\vdash_{PA} -\text{Bew}[\bot] \rightarrow -R$, then $\vdash_{PA} -\text{Bew}[\bot] \rightarrow (R \;\&\; -R)$, and $\vdash_{PA} \text{Bew}[\bot]$, which is absurd. So neither $\vdash_{PA} -\text{Bew}[\bot] \rightarrow -R$ nor $\vdash_{PA} -R \rightarrow -\text{Bew}[\bot]$.[2])

Despite these results, Rosser sentences are presently quite poorly understood. The question whether all Rosser sentences are equivalent was first asked by Kreisel, and is still open. [All Gödel sentences, fixed points of $-$Bew(x), are equivalent to $-$Bew[$\bot$]; $\top$ and $\bot$ are inequivalent fixed points of *Prov*(x).]

Since $\vdash_{PA} \text{Prov}(\ulcorner\bot\urcorner) \rightarrow \bot$, by (1), $\vdash_{PA} \text{Prov}(\ulcorner(\text{Prov}(\ulcorner\bot\urcorner) \rightarrow \bot)\urcorner)$. It is thus clear that $\nvdash_{PA} \text{Prov}(\ulcorner(\text{Prov}(\ulcorner\bot\urcorner) \rightarrow \bot)\urcorner) \rightarrow \text{Prov}(\ulcorner\bot\urcorner)$; otherwise, $\vdash_{PA} \bot$. Consequently, either for some S, S', $\nvdash_{PA}$ Prov(⌜$(S \rightarrow S')$⌝) $\rightarrow$ (Prov(⌜S⌝) $\rightarrow$ Prov(⌜S'⌝)), or for some S, $\nvdash_{PA}$ Prov(⌜S⌝) $\rightarrow$ Prov(⌜Prov(⌜S⌝)⌝) (cf. the proof of Theorem 1 in Chapter 3); but which is the case is apparently still an open question, as is the question what the truth-value is of Prov(Sub(**k**,**1**,**k**)), where k is the Gödel number of Prov(Sub(x_1,**1**,x_1)). The (awful? attractive?) possibility exists that the answers to these questions may depend on the way sentences of arithmetic are assigned Gödel numbers.

The argument of Rosser's that we gave, which showed that it is provable in arithmetic that if arithmetic is consistent then it is incomplete, can be used, and was originally used by Rosser, to show the incompleteness of a wide variety of theories.

For let T be a theory (in the language of arithmetic) with a recursive axiom system that is a consistent extension of Q (= Robinson's arithmetic, a very weak subtheory of PA). Then there is a formula $\mathrm{Pf}_T(y,x)$ such that for every n, S, if n is the Gödel number of a proof in T of S, then $\vdash_T \mathrm{Pf}_T(\mathbf{n},$⌜$S$⌝$)$, but if not, then $\vdash_T -\mathrm{Pf}_T(\mathbf{n},$⌜$S$⌝$)$. And for every predicate $P(x)$ of the language of T, there exists a sentence S such that $\vdash_T S \leftrightarrow P($⌜$S$⌝$)$. Let $\mathrm{Bew}_T(x)$ and $\mathrm{Prov}_T(x)$ be defined from $\mathrm{Pf}_T(y,x)$ in the obvious ways:

$$\begin{aligned}\mathrm{Bew}_T(x) &= \exists y \mathrm{Pf}_T(y,x);\\ \mathrm{Prov}_T(x) &= \exists y(\mathrm{Pf}_T(y,x)\\ &\quad \& \forall z(z < y \rightarrow -\mathrm{Pf}_T(z,\mathrm{Neg}(z)))).\end{aligned}$$

Rosser's argument shows that if $\vdash_T R \leftrightarrow -$Prov(⌜$R$⌝), then R is an undecidable sentence of T, and therefore that T is incomplete.

However, it is just not in general true that if $\vdash_T S \leftrightarrow -\mathrm{Bew}_T($⌜$S$⌝$)$, then S is an *undecidable* sentence of T. (What *is* true is that S is not a *provable* sentence of T.) For let PA^+ be the theory obtained by adjoining $\mathrm{Bew}_{PA}($⌜$\perp$⌝$)$ to PA. PA^+ is consistent and has a recursive axiom system, and is therefore incomplete. Let $\mathrm{Pf}_{\mathrm{PA+}}(y,x)$ be defined in a sufficiently reasonable way, so that $\vdash_{\mathrm{PA}} \forall yx(\mathrm{Pf}_{\mathrm{PA}}(y,x) \rightarrow \mathrm{Pf}_{\mathrm{PA+}}(y,x))$, whence $\vdash_{\mathrm{PA}} \forall x(\mathrm{Bew}_{\mathrm{PA}}(x) \rightarrow \mathrm{Bew}_{\mathrm{PA+}}(x))$, and so $\vdash_{\mathrm{PA+}} \forall x(\mathrm{Bew}_{\mathrm{PA}}(x) \rightarrow \mathrm{Bew}_{\mathrm{PA+}}(x))$. But then, since $\vdash_{\mathrm{PA}} \mathrm{Bew}_{\mathrm{PA}}(\ulcorner\bot\urcorner) \rightarrow \mathrm{Bew}_{\mathrm{PA}}(\ulcorner S\urcorner)$, $\vdash_{\mathrm{PA+}} \mathrm{Bew}_{\mathrm{PA}}(\ulcorner S\urcorner)$, and so $\vdash_{\mathrm{PA+}} \mathrm{Bew}_{\mathrm{PA+}}(\ulcorner S\urcorner)$, for *all* sentences S. Thus if $\vdash_{\mathrm{PA+}} S \leftrightarrow -\mathrm{Bew}_{\mathrm{PA+}}(\ulcorner S\urcorner)$, then $\vdash_{\mathrm{PA+}} -S$, and S is *disprovable* in PA^+ and not an undecidable sentence of PA^+. Rosser's method therefore shows incompleteness where Gödel's does not. (A version of PA^+ was first discussed by Gödel in "On formally undecidable propositions . . . ," who showed that it is consistent but ω-inconsistent. Gödel did not demonstrate that it is incomplete.)

Exercise

Show that $[\Box(\Diamond\top \rightarrow (-\Box p \ \& \ -\Box -p)) \ \& \ \Box(-\Box p \rightarrow p) \ \& \ \Diamond\Diamond\top] \rightarrow [\Box(\Diamond\top \rightarrow p) \ \& \ -\Box(p \rightarrow \Diamond\top) \ \& \ -\Box(-p \rightarrow \Diamond\top) \ \& \ -\Box(\Diamond\top \rightarrow -p)]$ is a theorem of G.

11

The fixed-point theorem

The present chapter is almost entirely devoted to a proof of the beautiful fixed-point theorem of de Jongh and Sambin.[1]

Theorem

Let $A(p,\mathsf{q})$ be modalized in p. Then there is a sentence $H(\mathsf{q})$ such that

$$\vdash_G H(\mathsf{q}) \leftrightarrow A(H(\mathsf{q}),\mathsf{q}) \text{ and}$$
$$\vdash_G \Box(p \leftrightarrow A(p,\mathsf{q})) \rightarrow \Box(p \leftrightarrow H(\mathsf{q})).$$

[Recall our notational conventions: '$A(p,\mathsf{q})$' denotes a sentence that contains no sentence letters except p,q; '$H(\mathsf{q})$' denotes a sentence that contains no sentence letters except q; and p is identical with none of the sentence letters in the sequence q. q may be the empty sequence; if it is, $H(\mathsf{q})$ is letterless. As always, $A(p,\mathsf{q})$ is *modalized in p* if every occurrence of p in $A(p,\mathsf{q})$ is in the scope of some occurrence of $\Box$. 'A' and 'H' abbreviate '$A(p,\mathsf{q})$' and '$H(\mathsf{q})$'.]

Here is the proof of the theorem:

Let $\mathsf{q} = q_1, \ldots, q_s$ (if q is nonempty).

Let n be the number of subsentences of A that are necessitations, that is, of the form $\Box D$. Let these subsentences be $\Box D_1, \ldots, \Box D_n$, and for each l, $1 \leqslant l \leqslant n$, let $C_l = \Box D_l$.

We now (simultaneously) define 'sentence of degree m' and 'm-character': The *sentences of degree* 0 are $\top$ and the sentences $q_1, \ldots, q_s$. If $E_1, \ldots, E_r$ are all the sentences of degree $\leqslant m$, then the *m-characters* are the 2^r sentences $E_1 \,\&\, E_2 \,\&\, \cdots \,\&\, E_r$, $-E_1 \,\&\, E_2 \,\&\, \cdots \,\&\, E_r$, $E_1 \,\&\, -E_2 \,\&\, \cdots \,\&\, E_r, \ldots$, and $-E_1 \,\&\, -E_2 \,\&\, \cdots \,\&\, -E_r$, and the *sentences of degree* $m + 1$ are then the 2^r sentences $\Diamond(E_1 \,\&\, E_2 \,\&\, \cdots \,\&\, E_r)$, $\Diamond(-E_1 \,\&\, E_2 \,\&\, \cdots \,\&\, E_r)$, $\Diamond(E_1 \,\&\, -E_2 \,\&\, \cdots \,\&\, E_r), \ldots$, and $\Diamond(-E_1 \,\&\, -E_2 \,\&\, \cdots \,\&\, -E_r)$.

We shall call a triple R, U, K *good* if for some m, $0 \leqslant m \leqslant n$, R is a conjunction of $n - m$ distinct subsentences of A of the form $\Box D$ (if $m = 0$, then all the C_is are conjuncts of R; if $m = n$, then R is the null conjunction $\bigwedge$), U is an m-character, and K is either p or a subsentence of A. m is the *index* of the triple.

In Chapter 9, we defined $\boxdot Y$ to be $(\Box Y \,\&\, Y)$. If $\vdash_G \boxdot Y \rightarrow Z$, then $\vdash_G \Box Y \rightarrow \Box Z$. And $\vdash_G \boxdot Y \rightarrow \Box(\boxdot Y)$.

We shall now assign a truth-value $\mathrm{X}(R, U, K)$ to each good triple R, U, K and then show that if $\mathrm{X}(R, U, K) = \top$, then $\vdash_G \boxdot(p \leftrightarrow A) \,\&\, R \,\&\, U \rightarrow K$, and if $\mathrm{X}(R, U, K) = \bot$, then $\vdash_G \boxdot(p \leftrightarrow A) \,\&\, R \,\&\, U \rightarrow -K$. It will turn out that the disjunction of all n-characters U such that $\mathrm{X}(\bigwedge, U, p) = \top$ is a satisfactory H.

We may suppose that X has been defined on all good triples of index $<m$, and that the index of R, U, K is m. We set

$\mathrm{X}(R, U, \bot) = \bot$;
$\mathrm{X}(R, U, q_k) = \top$ if q_k is a conjunct of U;
$\mathrm{X}(R, U, q_k) = \bot$ if q_k is not a conjunct of U;
$\mathrm{X}(R, U, \Box D) = \top$ if $\Box D$ is one of the conjuncts of R;
$\mathrm{X}(R, U, \Box D) = \top$ if $\Box D$ is not one of the conjuncts of R, but for every $(m - 1)$-character V such that $\Diamond V$ is one of the conjuncts of U, $\mathrm{X}(R \,\&\, \Box D, V, D) = \top$;

$X(R, U, \Box D) = \bot$ if $\Box D$ is not one of the conjuncts of R, and for some $(m-1)$-character V such that $\Diamond V$ is one of the conjuncts of U, $X(R \& \Box D, V, D) = \bot$

$X(R, U, K_1 \rightarrow K_2) = \top$ if either $X(R, U, K_1) = \bot$ or $X(R, U, K_2) = \top$;

$X(R, U, K_1 \rightarrow K_2) = \bot$ if both $X(R, U, K_1) = \top$ and $X(R, U, K_2) = \bot$; and

$X(R, U, p) = X(R, U, A)$.

For every k $(1 \leqslant k \leqslant s)$, exactly one of q_k, $-q_k$ is a conjunct of each m-character. And since p occurs in A only in the scope of $\Box$, every subsentence of A is a q_k, or a C_l, or a truth-functional combination of q_ks and C_ls, or A – which is itself a truth-functional combination of q_ks and C_ls – or p, or a truth-functional combination of p, q_ks, and C_ls. It follows that X is defined on every good triple R, U, K with index m and therefore that X is defined on all good triples.

Lemma 1

For every good triple, R, U, K, if $X(R, U, K) = \top$, then $\vdash_G \boxdot(p \leftrightarrow A) \& R \& U \rightarrow K$, and if $X(R, U, K) = \bot$, then $\vdash_G \boxdot(p \leftrightarrow A) \& R \& U \rightarrow -K$.

Proof. Induction on the index m of the triple R, U, K. Since $-q_k$ is a conjunct of U iff q_k is not a conjunct of U, and since $\vdash_G \boxdot(p \leftrightarrow A) \rightarrow (p \leftrightarrow A)$, there are only two nontrivial cases: (1) $m > 0$, $\Box D$ not a conjunct of R, and $X(R, U, \Box D) = \top$; and (2) $m > 0$, $\Box D$ not a conjunct of R, and $X(R, U, \Box D) = \bot$.

We defined the canonical model $\mathcal{M}_G$, $= \langle W_G, R_G, P_G \rangle$, in Chapter 7. The sentences valid in $\mathcal{M}_G$ are precisely the theorems of G. (In what follows we shall omit '$\mathcal{M}_G$' before '$\vDash$'.)

Case 1. Suppose (for *reductio*) that $\nvdash_G \boxdot(p \leftrightarrow A) \& R \& U \rightarrow \Box D$. Then for some w in W_G, $\nvDash_w \boxdot(p \leftrightarrow A)$

$\& R \& U \rightarrow \Box D$, and therefore $\models_w \boxdot(p \leftrightarrow A)$, $\models_w R$, $\models_w U$, and $\not\models_w \Box D$. Since $\vdash_G \Box(\Box D \rightarrow D) \rightarrow \Box D$, $\not\models_w \Box(\Box D \rightarrow D)$, and so for some x such that wR_Gx, $\not\models_x \Box D \rightarrow D$. Thus $\models_x \Box D$ and $\not\models_x D$. Since R is a conjunction of necessitations, $\vdash_G R \rightarrow \Box R$, and therefore $\models_w \Box R$, and $\models_x R$. And since $\vdash_G \boxdot(p \leftrightarrow A) \rightarrow \Box(\boxdot(p \leftrightarrow A))$, we also have $\models_x \boxdot(p \leftrightarrow A)$. Now, by the case assumption and the induction hypothesis, we have that for every $(m - 1)$-character V such that $\Diamond V$ is one of the conjuncts of U, $\vdash_G \boxdot(p \leftrightarrow A) \& R \& \Box D \& V \rightarrow D$. ($R \& \Box D, V, D$ has index $m - 1$.) Thus, for every $(m - 1)$-character V such that $\Diamond V$ is a conjunct of U, $\models_x \boxdot(p \leftrightarrow A) \& R \& \Box D \& V \rightarrow D$, and thus $\not\models_x V$. But if V is an $(m - 1)$-character such that $\Diamond V$ is not a conjunct of U, then $-\Diamond V$ is a conjunct of U, and so $\vdash_G U \rightarrow -\Diamond V$, whence $\models_w -\Diamond V$, $\models_w \Box - V$, and so $\not\models_x V$. Thus for *every* $(m - 1)$-character V, $\not\models_x V$. But the disjunction Y of all $(m - 1)$-characters is a theorem of the propositional calculus, and hence of G, and therefore $\models_x Y$. The contradiction shows our original supposition to be false.

Case 2. Since $X(R, U, \Box D) = \perp$, by the case assumption and the induction hypothesis, we have that for some $(m - 1)$-character V such that $\Diamond V$ is a conjunct of U,

(*) $\quad \vdash_G \boxdot(p \leftrightarrow A) \& R \& \Box D \& V \rightarrow -D.$

Suppose now that $\nvdash_G \boxdot(p \leftrightarrow A) \& R \& U \rightarrow -\Box D$. Then for some w in W_G, $\models_w \boxdot(p \leftrightarrow A)$, $\models_w R$, $\models_w U$, and $\models_w \Box D$. Since $\vdash_G U \rightarrow \Diamond V$, $\models_w \Diamond V$, and hence for some x such that wR_Gx, $\models_x V$, $\models_x \boxdot(p \leftrightarrow A)$ and $\models_x R$ (R is a conjunction of necessitations). But since $\vdash_G \Box D \rightarrow \Box(\Box D \& D)$, we also have that $\models_x \Box D$ and $\models_x D$, and thus our supposition contradicts (*). Lemma 1 is thus proved. ⊣

Now let H be the disjunction of all n-characters U such that $X(\wedge, U, p) = \top$. By Lemma 1, if $X(\wedge, U, p)$

$= \top$, then $\vdash_G \boxdot(p \leftrightarrow A)$ & $U \rightarrow p$, and thus $\vdash_G \boxdot(p \leftrightarrow A) \rightarrow (H \rightarrow p)$. But if $X(\bigwedge, U, p) = \bot$, then $\vdash_G \boxdot(p \leftrightarrow A)$ & $U \rightarrow -p$, and so $\vdash_G \boxdot(p \leftrightarrow A) \rightarrow (p \rightarrow -U)$. Thus if H' is the disjunction of all n-characters U such that $X(\bigwedge, U, p) = \bot$, then $\vdash_G \boxdot(p \leftrightarrow A) \rightarrow (p \rightarrow -H')$. But since for every n-character U, either $X(\bigwedge, U, p) = \top$ or $X(\bigwedge, U, p) = \bot$, we have that $\vdash_G H \vee H'$, and so $\vdash_G \boxdot(p \leftrightarrow A) \rightarrow (p \rightarrow H)$, whence $\vdash_G \boxdot(p \leftrightarrow A) \rightarrow (p \leftrightarrow H)$, and therefore $\vdash_G \Box(p \leftrightarrow A) \rightarrow \Box(p \leftrightarrow H)$, that is, $\vdash_G \Box(p \leftrightarrow A(p,\mathsf{q})) \rightarrow \Box(p \leftrightarrow H(\mathsf{q}))$.

We now show that $\vdash_G H(\mathsf{q}) \leftrightarrow A(H(\mathsf{q}),\mathsf{q})$.

Lemma 2
$\vdash_G \Box(p \leftrightarrow A(p,\mathsf{q})) \rightarrow (A(p,\mathsf{q}) \leftrightarrow A(H(\mathsf{q}),\mathsf{q}))$.

Proof. As $A(p,\mathsf{q})$ is modalized in p, there are sentences $A_1(p,\mathsf{q}), \ldots, A_n(p,\mathsf{q})$ such that $A(p,\mathsf{q})$ is composed from $\Box A_1(p,\mathsf{q}), \ldots, \Box A_n(p,\mathsf{q})$, and q by some truth-functional mode of composition and $A(H(\mathsf{q}),\mathsf{q})$ is composed from $\Box A_1(H(\mathsf{q}),\mathsf{q}), \ldots, \Box A_n(H(\mathsf{q}),\mathsf{q})$, and q by the same mode of composition. By Theorems 5 and 9 of Chapter 1, for each i, $1 \leqslant i \leqslant n$,

$$\vdash_G \Box(p \leftrightarrow H(\mathsf{q})) \rightarrow \Box(A_i(p,\mathsf{q}) \leftrightarrow A_i(H(\mathsf{q}),\mathsf{q})),$$

and then by normality,

$$\vdash_G \Box(p \leftrightarrow H(\mathsf{q})) \rightarrow (\Box A_i(p,\mathsf{q}) \leftrightarrow \Box A_i(H(\mathsf{q}),\mathsf{q})).$$

Thus by the propositional calculus,

$$\vdash_G \Box(p \leftrightarrow H(\mathsf{q})) \rightarrow (A(p,\mathsf{q}) \leftrightarrow A(H(\mathsf{q}),\mathsf{q})).$$

But, as we have seen,

$$\vdash_G \Box(p \leftrightarrow A(p,\mathsf{q})) \rightarrow \Box(p \leftrightarrow H(\mathsf{q})),$$

and therefore

$$\vdash_G \Box(p \leftrightarrow A(p,\mathsf{q})) \rightarrow (A(p,\mathsf{q}) \leftrightarrow A(H(\mathsf{q}),\mathsf{q})). \quad \dashv$$

Lemma 3
$\vdash_G \boxdot(p \leftrightarrow A(p,\mathsf{q})) \rightarrow (H(\mathsf{q}) \leftrightarrow A(H(\mathsf{q}),\mathsf{q}))$.

Proof. In the course of showing that $\vdash_G \Box(p \leftrightarrow A(p,\mathsf{q})) \rightarrow \Box(p \leftrightarrow H(\mathsf{q}))$, we showed that $\vdash_G \boxdot(p \leftrightarrow A) \rightarrow (p \leftrightarrow H)$, that is, $\vdash_G \boxdot(p \leftrightarrow A(p,\mathsf{q})) \rightarrow (p \leftrightarrow H(\mathsf{q}))$. But by the definition of '$\boxdot$', $\vdash_G \boxdot(p \leftrightarrow A(p,\mathsf{q})) \rightarrow (p \leftrightarrow A(p,\mathsf{q}))$ and $\vdash_G \boxdot(p \leftrightarrow A(p,\mathsf{q})) \rightarrow \Box(p \leftrightarrow A(p,\mathsf{q}))$, whence by Lemma 2, $\vdash_G \boxdot(p \leftrightarrow A(p,\mathsf{q})) \rightarrow (A(p,\mathsf{q}) \leftrightarrow A(H(\mathsf{q}),\mathsf{q}))$. Therefore, $\vdash_G \boxdot(p \leftrightarrow A(p,\mathsf{q})) \rightarrow (H(\mathsf{q}) \leftrightarrow A(H(\mathsf{q}),\mathsf{q}))$. ⊣

Lemma 4
Suppose that B is a sentence in which p does not occur and $\vdash_G \boxdot(p \leftrightarrow A(p,\mathsf{q})) \rightarrow B$. Then $\vdash_G B$.

Proof. Since $\vdash_G \boxdot(p \leftrightarrow A(p,\mathsf{q})) \rightarrow B$, by the propositional calculus and the definition of '$\boxdot$',

$$\vdash_G -B \rightarrow ([-(p \leftrightarrow A(p,\mathsf{q})) \;\&\; \Box(p \leftrightarrow A(p,\mathsf{q}))] \\ \mathrm{v}\; -\Box(p \leftrightarrow A(p,\mathsf{q}))).$$

And since $\vdash_G -\Box(p \leftrightarrow A(p,\mathsf{q})) \rightarrow \Diamond(-(p \leftrightarrow A(p,\mathsf{q})) \;\&\; \Box(p \leftrightarrow A(p,\mathsf{q})))$, by Lemma 2 and normality,

$$\vdash_G -B \rightarrow (-(p \leftrightarrow A(H(\mathsf{q}),\mathsf{q})) \\ \mathrm{v}\; \Diamond-(p \leftrightarrow A(H(\mathsf{q}),\mathsf{q}))).$$

But since G is normal, every substitution instance of a theorem is a theorem, and thus if we substitute $A(H(\mathsf{q}),\mathsf{q})$ for p in the last mentioned theorem, then, since p occurs in neither B nor $A(H(\mathsf{q}),\mathsf{q})$, we obtain $\vdash_G -B \rightarrow (-(A(H(\mathsf{q}),\mathsf{q}) \leftrightarrow A(H(\mathsf{q}),\mathsf{q})) \;\mathrm{v}\; \Diamond-(A(H(\mathsf{q}),\mathsf{q}) \leftrightarrow A(H(\mathsf{q}),\mathsf{q})))$, whence, since $\vdash_G \boxdot(A(H(\mathsf{q}),\mathsf{q}) \leftrightarrow A(H(\mathsf{q}),\mathsf{q}))$, $\vdash_G B$. ⊣

Since p does not occur in $H(\mathsf{q}) \leftrightarrow A(H(\mathsf{q}),\mathsf{q})$, it is immediate from Lemmas 3 and 4 that $\vdash_G H(\mathsf{q}) \leftrightarrow A(H(\mathsf{q}),\mathsf{q})$. The fixed-point theorem is thus completely proved.

In our proof of the fixed-point theorem, we have not made appeal to the completeness theorem for G and have cited only the most elementary facts about $\mathcal{M}_G$, the universal model for G. In fact, a proof of the fixed-point theorem can be given that contains no use at all of any semantic facts. All that is needed for such a proof is a semantics-free proof of Lemma 1, and we can readily obtain one from the semantical proof of Lemma 1 given above.

Alternative proof of Lemma 1. (The first paragraph is the same.)

Case 1. Let $V_1, \ldots, V_i, \ldots, V_e$ be *all* the $(m-1)$-characters. If $\Diamond V_i$ is not a conjunct of the m-character U, then $-\Diamond V_i$ is a conjunct of U, and so $\vdash_G U \rightarrow -\Diamond V_i$, whence $\vdash_G -(U \,\&\, \Diamond V_i)$, and so $\vdash_G -(U \,\&\, \Diamond(\boxdot(p \leftrightarrow A) \,\&\, R \,\&\, \Box D \,\&\, V_i \,\&\, -D))$. But if $\Diamond V_i$ is a conjunct of U, then, since $\mathbf{X}(R, U, \Box D) = \top$, the induction hypothesis entitles us to suppose that $\vdash_G \boxdot(p \leftrightarrow A) \,\&\, R \,\&\, \Box D \,\&\, V_i \rightarrow D$, whence

$$\vdash_G -(\boxdot(p \leftrightarrow A) \,\&\, R \,\&\, \Box D \,\&\, V_i \,\&\, -D),$$
$$\vdash_G \Box -(\boxdot(p \leftrightarrow A) \,\&\, R \,\&\, \Box D \,\&\, V_i \,\&\, -D),$$
$$\vdash_G -\Diamond(\boxdot(p \leftrightarrow A) \,\&\, R \,\&\, \Box D \,\&\, V_i \,\&\, -D),$$

and we therefore have that for *every* $(m-1)$-character V_i, $\vdash_G -(U \,\&\, \Diamond(\boxdot(p \leftrightarrow A) \,\&\, R \,\&\, \Box D \,\&\, V_i \,\&\, -D))$. So

$$\vdash_G -(U \,\&\, \Diamond(\boxdot(p \leftrightarrow A) \,\&\, R \,\&\, \Box D \,\&\, V_1 \,\&\, -D)) \,\&\, \cdots \,\&\, -(U \,\&\, \Diamond(\boxdot(p \leftrightarrow A) \,\&\, R \,\&\, \Box D \,\&\, V_e \,\&\, -D)),$$
$$\vdash_G -[(U \,\&\, \Diamond(\boxdot(p \leftrightarrow A) \,\&\, R \,\&\, \Box D \,\&\, V_1 \,\&\, -D)) \vee \cdots \vee (U \,\&\, \Diamond(\boxdot(p \leftrightarrow A) \,\&\, R \,\&\, \Box D \,\&\, V_e \,\&\, -D))],$$

whence, by the distributivity of & over v, the distributivity of $\Diamond$ over v, and the distributivity of & over v again, we obtain

$$\vdash_G -[U \,\&\, \Diamond(\boxdot(p \leftrightarrow A) \,\&\, R \,\&\, \Box D \,\&\, (V_1 \vee \cdots \vee V_e) \,\&\, -D)].$$

But since $\vdash_G V_1 \vee \cdots \vee V_e$, we have

$$\vdash_G -[U \,\&\, \Diamond(\boxdot(p \leftrightarrow A) \,\&\, R \,\&\, \Box D \,\&\, -D)],$$

whence

$$\vdash_G -[\boxdot(p \leftrightarrow A) \,\&\, R \,\&\, U \,\&\, \Diamond(\boxdot(p \leftrightarrow A) \,\&\, R \,\&\, \Box D \,\&\, -D)].$$

But now, since R is a conjunction of necessitations, $\vdash_G \boxdot L \,\&\, \Diamond M \rightarrow \Diamond(\boxdot L \,\&\, M)$, and $\vdash_G \Box L \,\&\, \Diamond M \rightarrow \Diamond(\Box L \,\&\, M)$, we have that $\vdash_G -[\boxdot(p \leftrightarrow A) \,\&\, R \,\&\, U \,\&\, \Diamond(\Box D \,\&\, -D)]$. But since $\vdash_G -\Box D \leftrightarrow \Diamond(\Box D \,\&\, -D)$, $\vdash_G -[\boxdot(p \leftrightarrow A) \,\&\, R \,\&\, U \,\&\, -\Box D)]$, whence $\vdash_G \boxdot(p \leftrightarrow A) \,\&\, R \,\&\, U \rightarrow \Box D$.

Case 2. $X(R, U, \Box D) = \perp$, and so by the induction hypothesis we may suppose that for some $(m - 1)$-character V such that $\Diamond V$ is a conjunct of U, $\vdash_G \boxdot(p \leftrightarrow A) \,\&\, R \,\&\, \Box D \,\&\, V \rightarrow -D$. But then

$$\vdash_G -(\boxdot(p \leftrightarrow A) \,\&\, R \,\&\, \Box D \,\&\, D \,\&\, V),$$
$$\vdash_G \Box-(\boxdot(p \leftrightarrow A) \,\&\, R \,\&\, \Box D \,\&\, D \,\&\, V), \text{ and}$$
$$(*) \quad \vdash_G -\Diamond(\boxdot(p \leftrightarrow A) \,\&\, R \,\&\, \Box D \,\&\, D \,\&\, V).$$

Since $\Diamond V$ is a conjunct of U, $\vdash_G U \rightarrow \Diamond V$, and so $\vdash_G \boxdot(p \leftrightarrow A) \,\&\, R \,\&\, \Box D \,\&\, U \rightarrow \Diamond V$, whence, since $\vdash_G \Box D \rightarrow \Box(\Box D \,\&\, D)$,

$$\vdash_G \boxdot(p \leftrightarrow A) \,\&\, R \,\&\, \Box D \,\&\, U \rightarrow \Diamond(\boxdot(p \leftrightarrow A) \,\&\, R \,\&\, \Box D \,\&\, D \,\&\, V),$$

and so, by (*), $\vdash_G \boxdot(p \leftrightarrow A) \,\&\, R \,\&\, U \rightarrow -\Box D$. ⊣

The fixed-point theorem is noteworthy not merely because it informs us of a pretty fact about some bizarre system of modal logic, but because it tells us something of great interest about provability, self-reference, and modes of sentence and predicate formation in formalized theories. What it says can be roughly expressed this way: There is a correspondence between certain modes of formation of predicates out of Bew(x), truth-functions, and sentences S and certain modes of formation of sentences out of the same materials, such that every sentence that can be described, "self-referentially," as equivalent to the assertion that it itself has the property expressed by a predicate constructed in one of the modes of predicate formation can also be described, "non-self-referentially," as equivalent to a sentence constructed in the corresponding mode of sentence formation; moreover, the correspondence is "uniform in" S. The modes of predicate formation can be taken to be the functions $\lambda\phi(A,\phi)(x)$(cf. Chapter 4), with A modalized in p (but possibly containing sentence letters other than p); the modes of sentence construction can be taken to be the functions $\lambda\phi K^{\phi}$, with K containing only letters contained in A that are different from p. The fixed-point theorem tells us that there is an effective function Ψ from sentences modalized in p to modal sentences, such that for every realization ϕ (notice the uniformity: Ψ does not depend on ϕ), the fixed points of the predicate $(A,\phi)(x)$ are precisely the sentences equivalent to $\Psi(A)^{\phi}$. $\lambda\phi\Psi(A)^{\phi}$ is thus the mode of sentence construction "corresponding to" $\lambda\phi(A,\phi)(x)$.

We began by noting that the Gödel sentences, the fixed points of $-\text{Bew}(x)$, are exactly the sentences equivalent to the consistency assertion, $-\text{Bew}[\bot]$ $(=(\Diamond\top)^{\phi})$. We observed other similar phenomena, and in Chapter 9 we determined that these phenomena were instances of a striking regularity: The Gödelian fixed points are the sentences equivalent to deictic sentences. But the Gödelian fixed points, as we saw in Chapter 4, are the fixed points of the predicates $(A,\phi)(x)$, with A modalized in p and containing no sentence letters other than p; and the deictic sentences are the sentences K^{ϕ}, with K letterless. That striking regularity, then, is itself an instance of the more encompassing regularity characterizing formalized theories of which the fixed-point theorem informs us.

Exercises

1 Let $A(p,q) =$ (a) $\Box p \rightarrow q$; (b) $\Box p \; \& \; q$. Find $H(q)$.

2 Formulate and prove a generalization of the fixed-point theorem about sentences $A_1(p_1, \ldots, p_n, \mathsf{q}), \ldots, A_n(p_1, \ldots, p_n, \mathsf{q})$ modalized in all of $p_1, \ldots, p_n$.

3 Show that if q is empty, then for every m-character U there is a member V of $\{\bot, \Diamond^m\top\} \cup \{-\Diamond^{i+1}\top \; \& \; \Diamond^i\top | i < m\}$ such that $\vdash_G U \leftrightarrow V$.

12

Solovay's completeness theorems

G, as we know, is sound and complete with respect to finite models with a transitive and well-capped accessibility relation; that is, all and only the theorems of G are valid in all models of this sort. We also know that all translations of all theorems of G are theorems of arithmetic. We are now going to prove the converse, Solovay's Completeness Theorem for G, according to which a modal sentence is a theorem of G if all of its translations are theorems of arithmetic. Solovay's Completeness Theorem for G thus tells us that if a modal sentence A is not a theorem of G, then there is a realization ϕ such that A^{ϕ} is not a theorem of arithmetic.

It would be a mistake to suppose that Solovay's Completeness Theorem for G is of interest merely because it informs us of the power of the modal calculus G. Interesting information about Peano Arithmetic can be inferred by means of Solovay's theorem from the knowledge that certain modal sentences are not theorems of G. For example, consider the sentence $A = \Box(\Box p \vee \Box -p) \rightarrow (\Box p \vee \Box -p)$. Although A is clearly a theorem of G*, it is not a theorem of G: Let $W = \{1,2,3\}$; $R = \{\langle 1,2\rangle, \langle 1,3\rangle\}$; and let $P(2,p) = \bot$ and $P(3,p) = \top$. Then $\langle W,R,P\rangle \nvDash_1 A$. Since $\langle W,R,P\rangle$ is transitive and well-capped, A is not a theorem of G. By Solovay's completeness theorem, for some ϕ, A^{ϕ} is not a theorem of PA. Let $S = \phi(p)$. Then $A^\phi = \text{Bew}[(\text{Bew}[S] \text{ v Bew}[-S])] \rightarrow (\text{Bew}[S] \text{ v Bew}[-S])$ is not a theorem of PA. Therefore, and perhaps surprisingly, there is a sentence S such that the following is consistent with arithmetic: S is undecidable and it is provable that S is decidable. Another application to PA of Solovay's completeness theorem is given in the next chapter.

If a modal sentence A is a theorem of G^*, then all translations of A are truths. Solovay's Completeness Theorem for G^* asserts that the converse also holds. We shall prove Solovay's Completeness Theorem for G^* after proving his completeness theorem for G. The proof of the completeness theorem for G^* will show the decidability of G^*, for it will show that a sentence A^* can be effectively associated with each modal sentence A so that $\vdash_G A^*$ iff $\vdash_{G*} A$. As we know, G is decidable, and it follows that G^* is decidable as well.

We begin the proof of Solovay's Completeness Theorem for G by appealing to the completeness theorem for G that was established in Chapters 7 and 8.

Suppose that the modal sentence A is not a theorem of G. Then by that earlier theorem, A is false at some world w of a model with a finite domain and a transitive and irreflexive accessibility relation. By Theorem 1 of Chapter 5, there are a finite set W containing w, a transitive and irreflexive relation R on W, and an evaluator P on W such that $W = \{w\} \cup \{x | wRx\}$ and $\langle W,R,P\rangle \nvDash_w A$. It is clear that we may assume that $w = 1$ and $W = \{1, \ldots, n\}$ without loss of generality.

Let $R' = R \cup \{\langle 0,j\rangle | 1 \leqslant j \leqslant n\}$.

Our first task is to construct sentences $S_0, S_1, \ldots, S_n$ of arithmetic such that for all i, j, $0 \leqslant i, j \leqslant n$:

(A) $\vdash_{PA} -(S_i \,\&\, S_j)$ if $i \neq j$;
(B) $\vdash_{PA} (S_0 \text{ v } S_1 \text{ v} \cdots \text{v } S_n)$;
(C) $\vdash_{PA} S_i \rightarrow -\text{Bew}[-S_j]$ if $iR'j$; and
(D) $\vdash_{PA} S_i \rightarrow \text{Bew}[-S_j]$ if $i \geqslant 1$ and not: $iR'j$.

Let $\text{Lh}(x_1)$ be a primitive recursive term for a function whose value for any number that represents a finite sequence is the "length" of the sequence, that is, the number of terms the sequence has.

Let $(x_1)_{x_2}$ be a primitive recursive term for a function whose value for any number g that represents a finite sequence and any number e is the eth term (if it exists) of the sequence represented by g ('g' goes with 'x_1'; 'e', with 'x_2').

Let F_m be the formula with Gödel number m.

Let $F_m b/\mathbf{j}$ be the result of substituting the numeral $\mathbf{j}$ for all free occurrences in F_m of the variable b.

Let $\Phi(x_1, x_2)$ be a primitive recursive term for a function whose value for each m, j is the Gödel number of the formula $-\exists c \forall a (a \geq c \rightarrow F_m b/\mathbf{j})$.

And let $B(y, a, b)$ be the following formula:

$$\exists s(\mathrm{Lh}(s) = a + \mathbf{1} \;\&\; (s)_0 = \mathbf{0} \;\&\; (s)_a = b \;\&$$

$$\forall x < a \bigwedge_{i:0 \leq i \leq n} \Bigl[(s)_x = \mathbf{i}$$

$$\rightarrow \Bigl\{\bigwedge_{j:iR'j} [\mathrm{Pf}(x, \Phi(y, \mathbf{j})) \rightarrow (s)_{x+1} = \mathbf{j}]$$

$$\& \Bigl[\Bigl\{\bigwedge_{j:iR'j} -\mathrm{Pf}(x, \Phi(y, \mathbf{j}))\Bigr\} \rightarrow (s)_{x+1} = (s)_x\Bigr]\Bigr\}\Bigr]\Bigr).$$

By the (generalized) diagonal lemma (see Chapter 3) there is a formula $G(a, b)$ with the variables a, b (and no others) free such that $\vdash_{\mathrm{PA}} G(a, b) \leftrightarrow B(\ulcorner G(a, b) \urcorner, a, b)$.

For each j, $0 \leq j \leq n$, let $S_j = \exists c \forall a (a \geq c \rightarrow G(a, \mathbf{j}))$. Then $\vdash_{\mathrm{PA}} \Phi(\ulcorner G(a, b) \urcorner, \mathbf{j}) = \ulcorner -\exists c \forall a (a \geq c \rightarrow G(a, \mathbf{j})) \urcorner = \ulcorner -S_j \urcorner$, and so

$$\text{(1)} \qquad \vdash_{\mathrm{PA}} G(a, b) \leftrightarrow \exists s(\mathrm{Lh}(s) = a + \mathbf{1} \;\&\; (s)_0 = \mathbf{0} \;\&\; (s)_a = b$$

$$\& \; \forall x < a \bigwedge_{i:0 \leq i \leq n} \Bigl[(s)_x = \mathbf{i}$$

$$\rightarrow \Bigl\{\bigwedge_{j:iR'j} [\mathrm{Pf}(x, \ulcorner -S_j \urcorner) \rightarrow (s)_{x+1} = \mathbf{j}]$$

$$\& \Bigl[\Bigl\{\bigwedge_{j:iR'j} -\mathrm{Pf}(x, \ulcorner -S_j \urcorner)\Bigr\} \rightarrow (s)_{x+1} = (s)_x\Bigr]\Bigr\}\Bigr]\Bigr).$$

Explanation. If $i \neq j$, then the sentence $-S_i$ is not identical to the sentence $-S_j$, and therefore there is no (number that is the Gödel number of a) proof of both $-S_i$ and $-S_j$. From (1) it can be seen that $G(a,b)$ defines a function h from the natural numbers into $\{0, 1, \ldots, n\}$ such that $h(0) = 0$ and for all natural numbers m, if $h(m) = i$, $iR'j$, and m is a proof of $-S_j$, then $h(m+1) = j$; but if $h(m) = i$ and m is not a proof of any sentence $-S_j$ with $iR'j$, then $h(m+1) = i$. But the definition of S_j tells us that S_j holds iff the limit of h is j, that is, iff the value of h is eventually constant and $= j$. Thus if $h(m) = i$, then $h(m+1) = j \neq h(m)$ iff $iR'j$ and m is the Gödel number of a proof that it is *not* the case that the limit of h is j. So if $h(m) = i$ and $iR'j$ for no j, then the limit of h is i; and in general if $h(m) = i$, then the limit of h is i or some j such that $iR'j$.

We now show that (A)–(D) hold. Since $\vdash_{PA} \exists! b G(a,b)$, (A) clearly holds.

(B): R' is well-capped. We first show by R'-induction that

(2) $$\vdash_{PA} G(a,\mathbf{i}) \rightarrow (S_i \vee \bigvee_{j:iR'j} S_j).$$

We may assume that for all j such that $iR'j$, $\vdash_{PA} G(c,\mathbf{j}) \rightarrow (S_j \vee \bigvee_{k:jR'k} S_k)$. From (1) we have that

$$\vdash_{PA} G(a,\mathbf{i}) \rightarrow \forall c \geqslant a(G(c,\mathbf{i}) \vee \bigvee_{j:iR'j} G(c,\mathbf{j})),$$

and thus

$$\vdash_{PA} G(a,\mathbf{i}) \rightarrow \forall c \geqslant a(S_i \vee \bigvee_{j:iR'j} G(c,\mathbf{j})),$$

which, together with the inductive assumption, yields

$$\vdash_{PA} G(a,\mathbf{i}) \rightarrow \forall c \geqslant a(S_i \vee \bigvee_{j:iR'j} (S_j \vee \bigvee_{k:jR'k} S_k)),$$

and therefore,

$$\vdash_{PA} G(a,\mathbf{i}) \rightarrow (S_i \vee \bigvee_{j:iR'j} (S_j \vee \bigvee_{k:jR'k} S_k)),$$

whence, as R' is transitive, (2) holds. But then, in particular, $\vdash_{PA} G(a, \mathbf{0}) \to (S_0 \vee S_1 \vee \cdots \vee S_n)$. And since $\vdash_{PA} G(\mathbf{0}, \mathbf{0})$, we have (B).

(C): Suppose that $iR'j$. The following argument can be formalized in PA: Suppose that the limit of h is i. Let m be the least number such that for all $r \geqslant m$, $h(r) = h(m) = i$. There are infinitely many proofs of each theorem of arithmetic. So if $-S_j$ is a theorem, then for some least $k > m$, k is the Gödel number of a proof of $-S_j$, and then $h(k+1) = j$, and so for some $r > m$, $h(r) = j \neq i$, a contradiction. Thus $-S_j$ is not a theorem. The formalization of this argument shows that $\vdash_{PA} S_i \to -\text{Bew}[-S_j]$, that is, that (C) holds.

(D): Suppose $i \geqslant 1$. Then $\vdash_{PA} G(a, \mathbf{i}) \to \exists x \text{Pf}(x, \ulcorner -S_i \urcorner)$. Since $\vdash_{PA} S_i \to \exists a G(a, \mathbf{i})$, we have $\vdash_{PA} S_i \not\to \text{Bew}[-S_i]$. Furthermore, if $i \neq j$ and not: $iR'j$, then by (2) and (A), $\vdash_{PA} G(a, \mathbf{i}) \to -S_j$, and so $\vdash_{PA} \text{Bew}[\exists a G(a, \mathbf{i})] \to \text{Bew}[-S_j]$. But it is clear from (1) that $G(a, b)$ is a Σ_1-formula. Consequently, $\exists a G(a, \mathbf{i})$ is a Σ_1-sentence, and so $\vdash_{PA} \exists a G(a, \mathbf{i}) \to \text{Bew}[\exists a G(a, \mathbf{i})]$ (cf. (v) of Chapter 2). Since $\vdash_{PA} S_i \to \exists a G(a, \mathbf{i})$, we have that $\vdash_{PA} S_i \to \text{Bew}[-S_j]$, and (D) holds.

(A)–(D) having been established, we now define ϕ. For each sentence letter p, let $\phi(p) = \bigvee\{S_i \mid (1 \leqslant i \leqslant n \;\&\; P(i,p) = \top) \vee (i = 0 \;\&\; P(1,p) = \top)\}$. [The function of the disjunct '$(i = 0 \;\&\; P(1,p) = \top)$' will become clear when we discuss G^*; were it not for G^*, this disjunct could have been omitted.]

Lemma 1

For all i, $1 \leqslant i \leqslant n$, and all subsentences B of A, if $\langle W, R, P\rangle \vDash_i B$, then $\vdash_{PA} S_i \to B^\phi$, and if $\langle W, R, P\rangle \nvDash_i B$, then $\vdash_{PA} S_i \to -B^\phi$.

Proof. Induction on the complexity of B. Suppose that $B = p$. (We henceforth omit '$\langle W, R, P\rangle$' before '$\vDash$'.) If $\vDash_i p$, then $P(i,p) = \top$, and thus $\vdash_{PA} S_i \to p^\phi$. But if $\nvDash_i p$, then $P(i,p) = \bot$; thus if $P(j,p) = \top$, then $i \neq j$, and since $i > 0$, by (A) we have $\vdash_{PA} S_i \rightarrow -S_j$ for *every* disjunct S_j of $\phi(p)$, and therefore $\vdash_{PA} S_i \rightarrow -p^\phi$. The truth-functional steps are perfectly straightforward. Suppose now that $B = \Box D$. Then $B^\phi = \text{Bew}[D^\phi]$. If $\models_i B$, then for all j such that iRj, $\models_j D$, and then by the induction hypothesis, for all j such that iRj, $\vdash_{PA} S_j \rightarrow D^\phi$. Then $\vdash_{PA} \left[\bigvee_{j:iRj} S_j\right] \rightarrow D^\phi$, and so $\vdash_{PA} \text{Bew}\left[\bigvee_{j:iRj} S_j\right] \rightarrow B^\phi$. But by (B), $\vdash_{PA} S_0 \text{ v } S_1 \text{ v } \cdots \text{ v } S_n$, and so $\vdash_{PA} \text{Bew}[S_0 \text{ v } S_1 \text{ v } \cdots \text{ v } S_n]$. Since $i \geqslant 1$, if not: iRj, then not: $iR'j$, and so by (D), $\vdash_{PA} S_i \rightarrow \text{Bew}[-S_j]$. We thus have $\vdash_{PA} S_i \rightarrow \text{Bew}\left[\bigvee_{j:iRj} S_j\right]$, and so $\vdash_{PA} S_i \rightarrow B^\phi$. Finally, if $\not\models_i B$, then for some j such that iRj, $\not\models_j D$. By the induction hypothesis, $\vdash_{PA} S_j \rightarrow -D^\phi$, whence $\vdash_{PA} -\text{Bew}[-S_j] \rightarrow -\text{Bew}[D^\phi]$. Since $iR'j$ if iRj, by (C) we have $\vdash_{PA} S_i \rightarrow -\text{Bew}[-S_j]$, and therefore $\vdash_{PA} S_i \rightarrow -B^\phi$. ⊣

It follows from Lemma 1 that $\vdash_{PA} S_1 \rightarrow -A^\phi$, and we thus have that $\vdash_{PA} -\text{Bew}[-S_1] \rightarrow -\text{Bew}[A^\phi]$. By (C), however, $\vdash_{PA} S_0 \rightarrow -\text{Bew}[-S_1]$, and therefore $\vdash_{PA} S_0 \rightarrow -\text{Bew}[A^\phi]$.

According to (A) and (B), exactly one of S_0, $S_1, \ldots, S_n$ is true. Suppose that $i \geqslant 1$. By (D), $\vdash_{PA} S_i \rightarrow \text{Bew}[-S_i]$. Thus if S_i is true, then $\text{Bew}[-S_i]$ is true, and then $-S_i$ is provable, and therefore S_i is not true. Thus if $i \geqslant 1$, S_i is not true. Therefore S_0 is true, $-\text{Bew}[A^\phi]$ is true, and A^ϕ is not a theorem of arithmetic. Solovay's Completeness Theorem for G is thus proved.

We now prove Solovay's Completeness Theorem for G^*. If A is a modal sentence, let

$$A^* = \left\{\bigwedge_{k=1}^{m} (\Box D_k \rightarrow D_k)\right\} \rightarrow A,$$

where $\Box D_1, \ldots, \Box D_m$ are all of the subsentences of A of the form $\Box D$.

Suppose now that A^* is not a theorem of G. Then for some $\langle\{1, \ldots, n\}, R, P\rangle$, with R transitive and irreflexive and $1Ri$ iff $1 < i \leqslant n$, we have that $\langle\{1, \ldots, n\}, R, P\rangle \nvDash_1 A^*$. Let $S_0, S_1, \ldots, S_n$ and ϕ be defined from $\langle\{1, \ldots, n\}, R, P\rangle$ as in the completeness proof for G.

Lemma 2
A^ϕ is false.

Proof. Since $\nvDash_1 A^*$, for all k, $1 \leqslant k \leqslant m$, $\vDash_1 \Box D_k \to D_k$ and $\nvDash_1 A$. We want to see that for all subsentences B of A, if $\vDash_1 B$, then $\vdash_{PA} S_0 \to B^\phi$, and if $\nvDash_1 B$, then $\vdash_{PA} S_0 \to -B^\phi$.

Suppose that $B = p$. If $\vDash_1 p$, then $P(1,p) = \top$, and so S_0 is one of the disjuncts of p^ϕ, and thus $\vdash_{PA} S_0 \to p^\phi$. But if $\nvDash_1 p$, then S_0 is not one of the disjuncts of p^ϕ, and so by (A), $\vdash_{PA} S_0 \to -p^\phi$.

The truth-functional steps are straightforward.

Suppose that $B = \Box D$. If $\vDash_1 \Box D$, then for all i, $1 < i \leqslant n$, $\vDash_i D$, and so by Lemma 1, $\vdash_{PA} S_i \to D^\phi$. But since $\vDash_1 \Box D \to D$, $\vDash_1 D$, and so by Lemma 1, $\vdash_{PA} S_1 \to D^\phi$, and by the hypothesis of the induction (D being simpler than B) $\vdash_{PA} S_0 \to D^\phi$. We thus have $\vdash_{PA} S_i \to D^\phi$ for all i, $0 \leqslant i \leqslant n$, and therefore by (B) $\vdash_{PA} D^\phi$, whence $\vdash_{PA} \mathrm{Bew}[D^\phi]$, that is, $\vdash_{PA} B^\phi$, and so $\vdash_{PA} S_0 \to B^\phi$. Lastly, if $\nvDash_1 \Box D$, then for some i, $1 < i \leqslant n$, $\nvDash_i D$, and so $\vdash_{PA} S_i \to -D^\phi$, whence $\vdash_{PA} -\mathrm{Bew}[-S_i] \to -\mathrm{Bew}[D^\phi]$. But since $0R'i$, by (C) we have $\vdash_{PA} S_0 \to -B^\phi$.

Thus since $\nvDash_1 A$, $\vdash_{PA} S_0 \to -A^\phi$. And since S_0 is true, A^ϕ is false. ⊣

We have thus shown that if A^* is not a theorem of G, then A^ϕ is false. Thus, if all translations of A are true, then A^* is a theorem of G. But if A^* is a theorem of G, then A^* is also a theorem of G^*, as are $A^* \to ((\Box D_1 \to D_1) \to (\cdots \to ((\Box D_m \to D_m) \to A) \cdots))$, which is a tautology, the consequent of that tautology, all of the sentences $\Box D_1 \rightarrow D_1, \ldots, \Box D_m \rightarrow D_m$, and A itself. Thus if all translations of A are true, then A is a theorem of G^*. Solovay's Completeness Theorem for G^* is thus proved.

We have now established that if all translations of A are true, then A^* is a theorem of G; if A^* is a theorem of G, then A is a theorem of G^*; and if A is a theorem of G^*, then all translations of A are true. Thus, for every modal sentence A, A is a theorem of G^* if and only if the sentence A^*, which can be effectively constructed from A, is a theorem of G. Since G is decidable, G^* is decidable too.

Exercise

Let ω-Con(x) be the natural formalization of 'is the Gödel number of a sentence that produces an ω-consistent theory when adjoined to PA'. Let '$A^{\omega\phi}$' be defined like 'A^{ϕ}' except that $(\Box B)^{\omega\phi} = -\omega\text{-Con}(\ulcorner -B^{\omega\phi}\urcorner)$. Show that $\vdash_G A$ iff for all ϕ, $\vdash_{PA} A^{\omega\phi}$, and $\vdash_{G*} A$ iff for all ϕ, $A^{\omega\phi}$ is true.

13

An $S4$-preserving proof-theoretical treatment of modality

Not every sentence of the form $\mathrm{Bew}(\ulcorner S \urcorner) \rightarrow S$ and, consequently, not every translation of $\Box p \rightarrow p$, is a theorem of arithmetic. Suppose, however, that we interpret the box as meaning 'it is provable and true that . . . ' rather than 'it is provable that . . . '. That is to say, suppose that we define a new notion of translation, the *truth-translation* ${}^{\phi}A$ of a modal sentence A under the realization ϕ, by setting

$$\begin{aligned} {}^{\phi}p &= \phi(p); \\ {}^{\phi}\bot &= \bot; \\ {}^{\phi}(A \rightarrow B) &= ({}^{\phi}A \rightarrow {}^{\phi}B); \text{ and} \\ {}^{\phi}(\Box A) &= (\mathrm{Bew}(\ulcorner {}^{\phi}A \urcorner) \ \& \ {}^{\phi}A). \end{aligned}$$

[Thus if ${}^{\phi}A = S$, then ${}^{\phi}(\Box A) = (\mathrm{Bew}(\ulcorner S \urcorner) \ \& \ S)$.] Then it is clear that every truth-translation of $\Box p \rightarrow p$ is a theorem of arithmetic, for every truth-translation of $\Box p \rightarrow p$ is a sentence $(\mathrm{Bew}(\ulcorner S \urcorner) \ \& \ S) \rightarrow S$ of arithmetic, and each of these is clearly a theorem.

[We cannot define the notion of *true sentence* by a predicate of arithmetic, but for each sentence S of arithmetic, we may take the arithmetization of the assertion that S is true to be S itself. If $\phi(p) = S$, ${}^{\phi}(\Box p)$ will then assert that S is provable and true.]

It is easily seen, moreover, that, for all S and S',

$$((\mathrm{Bew}(\ulcorner S \to S' \urcorner) \,\&\, (S \to S')) \to ((\mathrm{Bew}(\ulcorner S \urcorner) \,\&\, S) \to (\mathrm{Bew}(\ulcorner S' \urcorner) \,\&\, S'))$$ and

$$(\mathrm{Bew}(\ulcorner S \urcorner) \,\&\, S) \to (\mathrm{Bew}(\ulcorner (\mathrm{Bew}(\ulcorner S \urcorner) \,\&\, S) \urcorner) \,\&\, (\mathrm{Bew}(\ulcorner S \urcorner) \,\&\, S))$$

are theorems of arithmetic, and that if S is a theorem, then $(\mathrm{Bew}(\ulcorner S \urcorner) \,\&\, S)$ is a theorem too. It follows that every truth-translation of every theorem of $S4$ is a theorem of arithmetic. We know what the modal sentences are all of whose translations are theorems of arithmetic. The aim of this chapter is to characterize those modal sentences all of whose truth-translations are theorems of arithmetic.

The axioms of *the calculus S4Grz*[1] are those of $S4$ and the sentences of the form

$$\Box(\Box(A \to \Box A) \to A) \to A;$$

its rules of inference are modus ponens and necessitation.

It is easy to see that $\Box(\Box(p \to \Box p) \to p) \to p$ is not even a theorem of $S5$, and thus that the theorems of $S4Grz$ properly include those of $S4$. Let $W = \{0,1\}$; let $R = \{\langle 0,0\rangle,\langle 0,1\rangle,\langle 1,0\rangle,\langle 1,1\rangle\}$; and let $P(0,p) = \bot$ and $P(1,p) = \top$. R is an equivalence relation on W, and $\langle W,R,P\rangle$ is thus appropriate to $S5$. But $\nvDash_0 p$, $\nvDash_1 \Box p$, $\vDash_1 p$, $\nvDash_1 p \to \Box p$, $\nvDash_0 \Box(p \to \Box p)$, $\vDash_0 \Box(p \to \Box p) \to p$, $\vDash_1 \Box(p \to \Box p) \to p$, $\vDash_0 \Box(\Box(p \to \Box p) \to p)$, and $\nvDash_0 \Box(\Box(p \to \Box p) \to p) \to p$, and so $\nvDash_{S5} \Box(\Box(p \to \Box p) \to p) \to p$. ('$\langle W,R,P\rangle$' is omitted.)

R is said to be an *antisymmetric* relation if for all x,y, if xRy and yRx, then $x = y$. Frames $\langle W,R\rangle$ in which W is finite and R is a transitive and antisymmetric relation that is reflexive on W are called *finite weak partial orderings*.

We define a mapping ^{t} from modal sentences A to modal sentences ${}^tA$ by setting

$${}^tp = p;$$
$${}^t\bot = \bot;$$
$${}^t(A \to B) = ({}^tA \to {}^tB); \text{ and}$$
$${}^t(\Box A) = (\Box {}^tA \,\&\, {}^tA).$$

We shall show that if a modal sentence A satisfies one of the following four conditions, then it satisfies all the others:

(1) $\vdash_{S4Grz} A$;
(2) A is valid in all finite weak partial orderings;
(3) $\vdash_G {}^tA$; and
(4) every truth-translation of A is a theorem of arithmetic.

The equivalence of (1) and (2) is due to Segerberg.[2] The equivalance of these to (3) and to (4) was first observed by Goldblatt.[3] We shall show that (2) implies (1), (3) implies (2), (4) implies (3), and (1) implies (4), in that order.

The proof that (2) implies (1) follows much the same course as the completeness proof for G given in Chapter 7. We begin by supposing that A is not a theorem of S4Grz. Then there is a world t in W_{S4Grz}, the domain of the canonical model $\mathcal{M}_{S4Grz}$ for S4Grz, such that $\mathcal{M}_{S4Grz} \nvDash_t A$. We define W_1, R_1, P_1, $\mathcal{M}_1$, $\equiv$, w^o, W_2, R_2, P_2, $\mathcal{M}_2$, and $\sim$ from t, A, and $\mathcal{M}_{S4Grz}$ exactly as W_1, etc., were defined from t, A, and $\mathcal{M}_G$ in Chapter 7. W_2 is again finite, R_2 is again transitive, and, again, if wR_1x, then $w^oR_2x^o$. But now R_2 is reflexive on W_2: For if $w^o \in W_2$, then, since R_{S4Grz} is reflexive, wR_1w, and so $w^oR_2w^o$. We prove Lemmas 1 and 2 exactly as they were proved in Chapter 7.

Lemma 1
For every truth-functional combination B of subsentences of A and every w in W_1, $\mathcal{M}_1 \vDash_w B$ if and only if $\mathcal{M}_2 \vDash_{w^o} B$.

Lemma 2
For every subset S of W_2 there is a truth-functional combination C of subsentences of A such that for every u in W_1, $\mathcal{M}_1 \vDash_u C$ iff $u^o \in S$.

But before we can define R_3, we must prove an analogue of Lemma 3 of Chapter 7:

Lemma 3
If S is a nonempty subset of W_2, then for some a in W_1, $a^o \in S$ and for every x in W_1 such that aR_1x and $x^o \in S$, $x^o = a^o$.

Proof. Let S be a nonempty subset of W_2.

We define a sequence $a_0, a_1, \ldots$ of members of W_1 and a sequence $S_0, S_1, \ldots$ of subsets of W_2:

Let a_0 be any member of some $\equiv$-equivalence class in S (S is a nonempty set of $\equiv$-equivalence classes), and let $S_0 = S$.

Suppose a_i and S_i are defined. Define a_{i+1} and S_{i+1} as follows:

Case 1. There exists a b such that a_iR_1b, $b^o \in S_i$, and for some z^o in S_i, $z^o \notin \{x^o | bR_1x\}$. Then let a_{i+1} be any such b and let $S_{i+1} = S_i \cap \{x^o | a_{i+1}R_ix\}$ (which will be a proper subset of S_i).

Case 2. No such b exists. Then let $a_{i+1} = a_i$ and $S_{i+1} = S_i$.

It is evident that $S_0 \supseteq S_1 \supseteq \cdots$. Since S_0 is a subset of W_2, which is finite, it follows that for some i, $S_i = S_{i+1}$. Let m be the least such i and let $a = a_m$.

For all i, $a_i^o \in S$, and hence $a^o \in S$; we must show that for all x, if aR_1x and $x^o \in S$, then $x^o = a^o$.

Since R_1 is reflexive and transitive, $a_iR_1a_j$ whenever $i \leqslant j$. It follows by induction that if a_iR_1x and $x^o \in S$, then $x^o \in S_i$: For $S_0 = S$; and if $a_{i+1}R_1x$ and $x^o \in S$, then since $a_iR_1a_{i+1}$, a_iR_1x, whence by the induction hypothesis, $x^o \in S_i$, and therefore $x^o \in S_{i+1}$, as $a_{i+1}R_ix$.

Thus if aR_1x and $x^o \in S$, then $x^o \in S_m$. And since R_1 is reflexive and $a^o \in S$, $a^o \in S_m$. To complete the proof it thus suffices to show that S_m contains no member other than a^o. So suppose that $x^o \in S_m$ and $x^o \neq a^o$.

Since $x^o \neq a^o$, for some subsentence B of A, or the negation of one, $B \notin x$ and $B \in a$, whence $\mathcal{M}_1 \nvDash_x B$ and $\mathcal{M}_1 \vDash_a B$, and then by Lemma 1, $\mathcal{M}_2 \nvDash_{x^o} B$ and $\mathcal{M}_2 \vDash_{a^o} B$, and by Lemma 1 again $\mathcal{M}_1 \nvDash_y B$ for all y in x^o and $\mathcal{M}_1 \vDash_b B$ for all b in a^o.

(We omit '$\mathcal{M}_1$' before '$\vDash$' in what follows.)

By Lemma 2 (with S_m as S) there exists a truth-functional combination C of subsentences of A such that for all u in W_1, $\vDash_u C$ iff $u^o \in S_m$.

Now since $S_{m+1} = S_m$, aR_1a, $a^o \in S_m$, and $x^o \in S_m$, it follows from the definition of S_{m+1} that $x^o \in \{y^o|aR_1y\}$, and thus for some y, aR_1y and $y^o = x^o$. So $y \in x^o$ and thus $\nvDash_y B$. Since $y^o \in S_m$, $\vDash_y C$. So $\nvDash_y C \to B$ and then by the special axiom for $S4Grz$, $\nvDash_y \Box(\Box((C \to B) \to \Box(C \to B)) \to (C \to B))$. Thus for some b, yR_1b, $\vDash_b \Box((C \to B) \to \Box(C \to B))$ and $\vDash_b C$, whence $b^o \in S_m$. By transitivity, aR_1b. Since $S_{m+1} = S_m$, for some c, bR_1c and $c^o = a^o$, whence $c^o \in S_m$. Since bR_1c, $\vDash_c (C \to B) \to \Box(C \to B)$. Since $c^o = a^o$, $c \in a^o$, $\vDash_c B$, $\vDash_c C \to B$, and $\vDash_c \Box(C \to B)$. By transitivity, aR_1c. Since $S_{m+1} = S_m$, for some d, cR_1d and $d^o = x^o$. Since cR_1d, $\vDash_d C \to B$. Since $d^o \in S_m$, $\vDash_d C$, and so $\vDash_d B$. But since $d^o = x^o$, $d \in x^o$, and so $\nvDash_d B$, which is a contradiction. ⊣

A relation L is *connected on* a set S if for every x,y in S, either xLy or $x = y$ or yLx. L is called a *weak linear ordering of* S if L is a transitive, antisymmetric relation on S that is reflexive and connected on S.

Suppose now that S is a nonempty subset of W_2. By Lemma 3, let a be some element of W_1 that satisfies the condition given in Lemma 3, and let L_S be a weak linear ordering of S *whose last element is* a^o.

We can now define R_3. Let $R_3 = \{\langle w^o,x^o\rangle|$either $w^o \not\sim x^o$ and $w^oR_2x^o$ or $w^o \sim x^o$ and $w^oL_Sx^o$, where $S = \{y^o|w^o \sim y^o\}\}$.

R_3 is again a transitive subrelation of R_2. But now, since L_S is reflexive on S, R_3 is also reflexive on W_2.

And R_3 is now antisymmetric: For if $w^o R_3 x^o R_3 w^o$, then $w^o R_2 x^o R_2 w^o$, and so $w^o \sim x^o$, whence $w^o L_S x^o L_S w^o$, and so by the antisymmetry of L_S, $w^o = x^o$.

Thus $\langle W_2, R_3 \rangle$ is a finite weak partial ordering.

Let $\mathcal{M}_3 = \langle W_2, R_3, P_2 \rangle$.

Lemma 4

For every subsentence B of A and every w^o in W_2, $\mathcal{M}_2 \models_{w^o} B$ if and only if $\mathcal{M}_3 \models_{w^o} B$.

Proof. Induction on the complexity of B. As in the proof of Lemma 4 of Chapter 7, the only difficulty occurs in showing that if $\mathcal{M}_2 \not\models_{w^o} \Box D$, then $\mathcal{M}_3 \not\models_{w^o} \Box D$. So suppose that $\mathcal{M}_2 \not\models_{w^o} \Box D$.

By Lemma 1, $\mathcal{M}_1 \not\models_w \Box D$, and so $\Box D \notin w$. Let $S = \{y^o | w^o \sim y^o\}$, and let z^o be the L_S-last element of S. Then for some a in z^o, for every x such that aR_1x and $x^o \in S$, $x^o = a^o$. Since $a \in z^o$, $a^o = z^o$, and so $w^o L_S a^o$ and $w^o \sim a^o$. Since $w^o \sim a^o$ and $\Box D \notin w$, $\Box D \notin a$, and $\mathcal{M}_1 \not\models_a \Box D$. So for some b such that aR_1b, $\mathcal{M}_1 \not\models_b D$. By Lemma 1, $\mathcal{M}_2 \not\models_{b^o} D$, and then by the induction hypothesis, $\mathcal{M}_3 \not\models_{b^o} D$. We shall show that $w^o R_3 b^o$, for then $\mathcal{M}_3 \not\models_{w^o} \Box D$. As aR_1b, $a^o R_2 b^o$, and as $w^o \sim a^o$, $w^o R_2 b^o$. Thus if $w^o \not\sim b^o$, then $w^o R_3 b^o$. So suppose that $w^o \sim b^o$. Then since aR_1b and $b^o \in S$, $b^o = a^o$, and so $w^o L_S b^o$, whence $w^o R_3 b^o$. ⊣

As in Chapter 7, then, $\mathcal{M}_3 \not\models_{t^o} A$. And therefore A is invalid in some finite weak partial ordering, namely, $\langle W_2, R_3 \rangle$. Thus (2) implies (1).

We next show that (3) implies (2). For every relation R, let $I(R) = R - \{\langle x,x \rangle | xRx\}$. It is a perfectly mechanical matter to verify that if $\langle W, R \rangle$ is a finite weak partial ordering, then $\langle W, I(R) \rangle$ is a finite strict partial ordering.

Lemma 5
Let $\langle W, R\rangle$ be a finite weak partial ordering. Then for every sentence A, $\langle W, R, P\rangle \vDash_w A$ iff $\langle W, I(R), P\rangle \vDash_w {}^tA$.

Proof. Induction on the complexity of A. The only nontrivial case is that in which $A = \Box D$. But in this case $\langle W, R, P\rangle \vDash_w A$ iff for all x such that wRx, $\langle W, R, P\rangle \vDash_x D$, iff for all x such that $wI(R)x$ or $w = x$, $\langle W, R, P\rangle \vDash_x D$, iff (by the induction hypothesis) for all x such that $wI(R)x$ or $w = x$, $\langle W, I(R), P\rangle \vDash_x {}^tD$, iff $\langle W, I(R), P\rangle \vDash_w \Box {}^tD$ and $\langle W, I(R), P\rangle \vDash_w {}^tD$, iff $\langle W, I(R), P\rangle \vDash_w (\Box {}^tD \mathbin{\&} {}^tD)$, iff $\langle W, I(R), P\rangle \vDash_w {}^tA$. ⊣

Suppose now that $\vdash_G {}^tA$, and that $\langle W, R\rangle$ is a finite weak partial ordering. Let P be an evaluator on W and let w be a member of W. We must show that $\langle W, R, P\rangle \vDash_w A$. Since $\langle W, R\rangle$ is a finite weak partial ordering, $\langle W, I(R)\rangle$ is a finite strict partial ordering, and therefore by the completeness theorem for G established in Chapter 7, $\langle W, I(R), P\rangle \vDash_w {}^tA$. By Lemma 5, then, $\langle W, R, P\rangle \vDash_w A$. Thus (3) implies (2).

Via Lemma 6, we can prove that (4) implies (3).

Lemma 6
For all modal sentences A, ${}^\phi A = ({}^tA)^\phi$.

Proof. Again, the only slightly nontrivial case is that in which $A = \Box D$. But if $A = \Box D$, then ${}^\phi A = (\mathrm{Bew}(\ulcorner {}^\phi D \urcorner) \mathbin{\&} {}^\phi D) \underset{\text{i.h.}}{=} (\mathrm{Bew}(\ulcorner ({}^tD)^\phi \urcorner) \mathbin{\&} ({}^tD)^\phi) = ((\Box {}^tD)^\phi \mathbin{\&} ({}^tD)^\phi)$
$= (\Box {}^tD \mathbin{\&} {}^tD)^\phi = ({}^tA)^\phi$. ⊣

Now suppose that every truth-translation of A is a theorem of arithmetic, that is, for every realization ϕ, $\vdash_{\mathrm{PA}} {}^\phi A$. Then by Lemma 6, for every ϕ, $\vdash_{\mathrm{PA}} ({}^tA)^\phi$. By Solovay's Completeness Theorem for G, $\vdash_G {}^tA$. Thus (4) implies (3).

We now close the circle by showing that (1) implies (4). It is possible to give a direct proof of this implication by induction on proofs in *S4Grz* and appeal to Löb's theorem and the Hilbert-Bernays derivability conditions. We shall give a semantical proof instead.

Suppose, then, that for some realization ϕ, $\nvdash_{PA} {}^{\phi}A$. By Lemma 6, $\nvdash_{PA} ({}^{t}A)^{\phi}$, and therefore $\nvdash_{G} {}^{t}A$. Thus there is a finite strict partial ordering $\langle W,I\rangle$ in which ${}^{t}A$ is invalid. So for some evaluator P on W and some w in W, $\langle W,I,P\rangle \nvDash_{w} {}^{t}A$. Let $R = I \cup \{\langle x,x\rangle | x \in W\}$. Then $\langle W,R\rangle$ is a finite weak partial ordering, and $I(R) = I$. By Lemma 5, $\langle W,R,P\rangle \nvDash_{w} A$. The following lemma, a soundness theorem for *S4Grz*, will thus complete the demonstration of the equivalence of our four conditions.

Lemma 7

Every theorem of *S4Grz* is valid in every finite weak partial ordering.

Proof. We need only show that every sentence of the form $\Box(\Box(A \to \Box A) \to A) \to A$ is valid in every finite weak partial ordering; the reflexivity and transitivity of weak partial orderings take care of the other axioms.

Suppose that R is a transitive and reflexive relation on a finite set W such that for some $P, w, \langle W,R,P\rangle \nvDash_{w} \Box(\Box(A \to \Box A) \to A) \to A$. We shall show that R is not antisymmetric. We first define a sequence $w_0, w_1, \ldots$ of elements of W such that for all i, $w_i \neq w_{i+1}$ and $w_i R w_{i+1}$. By transitivity we shall have that $w_i R w_j$ if $i < j$.

Let $w_0 = w$. Then $\vDash_{w_0} \Box(\Box(A \to \Box A) \to A)$ and $\nvDash_{w_0} A$. Suppose that $\vDash_{w_{2n}} \Box(\Box(A \to \Box A) \to A)$ and $\nvDash_{w_{2n}} A$. By reflexivity, $\vDash_{w_{2n}} \Box(A \to \Box A) \to A$, and so $\nvDash_{w_{2n}} \Box(A \to \Box A)$. So for some w_{2n+1} such that $w_{2n} R w_{2n+1}$, $\nvDash_{w_{2n+1}} A \to \Box A$, $\vDash_{w_{2n+1}} A$, and $\nvDash_{w_{2n+1}} \Box A$. Since $\nvDash_{w_{2n}} A$ and $\vDash_{w_{2n+1}} A$, $w_{2n} \neq w_{2n+1}$. And since $\nvDash_{w_{2n+1}} \Box A$, for some w_{2n+2} such that $w_{2n+1} R w_{2n+2}$, $\nvDash_{w_{2n+2}} A$, whence $w_{2n+1} \neq w_{2n+2}$. By transitivity, $\vDash_{w_{2n+2}} \Box(\Box(A \to \Box A) \to A)$, and thus a sequence of the desired type exists.

Since W is finite, for some i and some $j > i$, $w_j = w_i$. Since $w_i \neq w_{i+1}$, $j > i + 1$. Thus $w_i R w_{i+1} R w_j$. R is therefore not antisymmetric. ⊣

S4Grz is decidable, for every nontheorem of *S4Grz* is invalid in some *finite* weak partial ordering.

14

The Craig Interpolation Lemma for G

We are going to prove the Craig Interpolation Lemma for G: If $\vdash_G A \to C$, then for some sentence B that contains no sentence letters not contained in both A and C, $\vdash_G A \to B$ and $\vdash_G B \to C$.[1] The Beth Definability Theorem for G (see below) follows in the usual way from the Craig Interpolation Lemma, and after proving the Beth Definability Theorem, we shall deduce from it the proof-theoretically more interesting half of the fixed-point theorem (which we proved on pp. 141–5). Some of the ideas in our proof of the Craig Interpolation Lemma for G were introduced by Solovay in "Provability interpretations of modal logic".[2]

We begin the proof of the Craig lemma with some definitions.

Let S (U) be the set of subsentences of A (C) that are either sentence letters or necessitations.

Let 's' and 's'' ('u' and 'u'') be variables ranging over functions from S (U) to $\{0, 1\}$. (We identify 0 with $\bot$, 1 with $\top$.)

Let $\Phi(s) = \bigwedge\{D' | D \in S\}$, where $D' = D$ if $s(D) = 1$ and $D' = -D$ if $s(D) = 0$. Let $\Phi(u) = \bigwedge\{D' | D \in U\}$, where $D' = D$ if $u(D) = 1$ and $D' = -D$ if $u(D) = 0$. [Note that every truth-functional combination of subsentences of A (C) is also a truth-functional combination of members of S (U) and is therefore equivalent in the propositional calculus to a (possibly empty) disjunction of sentences $\Phi(s)$ ($\Phi(u)$).]

Let $V_{su} = \{E \mid s(\Box E) = 0 \text{ or } u(\Box E) = 0\}$.

Lemma 1

Let $\Box D_0, \ldots, \Box D_{n-1}$ be all the necessitations $\Box D$ in $S \cup U$ such that either $s(\Box D) = 1$ or $u(\Box D) = 1$. Then if $\vdash_G -(\Phi(s) \;\&\; \Phi(u))$, then either (I) for some sentence letter p, $s(p) = 1 - u(p)$ or (II) for some sentence E in V_{su}, $\vdash_G -(-E \;\&\; \boxdot D_0 \;\&\; \cdots \;\&\; \boxdot D_{n-1})$.

Proof. Suppose that (I) and (II) are false. We shall show that $\nvdash_G -(\Phi(s) \;\&\; \Phi(u))$.

As (II) is false, for each sentence E in V_{su}, we let W_E, R_E, P_E, and w_E be such that R_E is a transitive, well-capped relation on W_E, $W_E = \{w_E\} \cup \{x \mid w_E R_E x\}$, and $\langle W_E, R_E, P_E\rangle \vDash_{w_E} (-E \;\&\; \boxdot D_0 \;\&\; \cdots \;\&\; \boxdot D_{n-1})$. Without loss of generality, we may assume that if $E_1, E_2 \in V_{su}$, $E_1 \neq E_2$, then W_{E_1} and W_{E_2} are disjoint.

Let w be some object not in any W_E.

Let $W = \{w\} \cup \bigcup \{W_E \mid E \in V_{su}\}$.

Let $R = \{\langle w,x\rangle \mid x \in W \;\&\; x \neq w\} \cup \bigcup \{R_E \mid E \in V_{su}\}$.

Define P by

$$\begin{aligned} P(w,p) &= s(p) \text{ if } p \in S; \\ P(w,p) &= u(p) \text{ if } p \in U; \\ P(w,p) &= \bot \text{ (arbitrarily) if } p \notin S \cup U; \text{ and} \\ P(x,p) &= P_E(x,p) \text{ if } x \in W_E. \end{aligned}$$

Since (I) is false, if $p \in S \cap U$, then $s(p) = u(p)$, and thus P assigns a unique truth-value to each pair consisting of a member of W and a sentence letter.

We now show that $\langle W,R,P\rangle \vDash_w \Phi(s) \;\&\; \Phi(u)$. Since R is a transitive, well-capped relation on W, it will follow that $\nvdash_G -(\Phi(s) \;\&\; \Phi(u))$. We shall show that $\langle W,R,P\rangle \vDash_w \Phi(s)$; the argument for $\Phi(u)$ is similar.

If p is a conjunct of $\Phi(s)$, then $s(p) = 1$, $P(w,p) = 1$, and $\langle W,R,P\rangle \vDash_w p$; if $-p$ is a conjunct of $\Phi(s)$, then $s(p)$

$= 0$, $P(w,p) = 0$, and $\langle W,R,P\rangle \vDash_w -p$. If $-\Box F$ is a conjunct of $\Phi(s)$, then $s(\Box F) = 0$, and so $\langle W_F,R_F,P_F\rangle \vDash_{w_F} -F$, whence by Theorem 1 of Chapter 5, $\langle W,R,P\rangle \vDash_{w_F} -F$. and so $\langle W,R,P\rangle \vDash_w -\Box F$, as wRw_F.

Finally, assume that $\Box F$ is a conjunct of $\Phi(s)$ and thus that $s(\Box F) = 1$. Suppose that wRx. We must show that $\langle W,R,P\rangle \vDash_x F$. Since wRx, for some E in V_{su}, $x \in W_E$. $\langle W_E,R_E,P_E\rangle \vDash_{w_E} \Box F$, and since $x \in W_E$, either $x = w_E$ or $w_E R_E x$. In either case, $\langle W_E,R_E,P_E\rangle \vDash_x F$, and then by Theorem 1 of Chapter 5, $\langle W,R,P\rangle \vDash_x F$.

Thus $\langle W,R,P\rangle \vDash_w \Phi(s)$. ⊣

Let 'X' ('Z') be a variable ranging over disjunctions of sentences $\Phi(s)$ ($\Phi(u)$).

Let $\mathscr{L}$ be the set of sentences containing no sentence letters not contained in both A and C.

Lemma 2

Suppose that for every pair s, u such that $\Phi(s)$ is a disjunct of X and $\Phi(u)$ is a disjunct of Z, there is a sentence Y in $\mathscr{L}$ such that $\vdash_G \Phi(s) \to Y$ and $\vdash_G \Phi(u) \to -Y$. Then for some sentence Y in $\mathscr{L}$, $\vdash_G X \to Y$ and $\vdash_G Z \to -Y$.

Proof. For every pair s, u such that $\Phi(s)$ is a disjunct of X and $\Phi(u)$ is a disjunct of Z, let Y_{su} be a sentence in $\mathscr{L}$ such that $\vdash_G \Phi(s) \to Y_{su}$ and $\vdash_G \Phi(u) \to -Y_{su}$. For each s such that $\Phi(s)$ is a disjunct of X, let $Y_s = \bigwedge\{Y_{su}|\Phi(u)$ is a disjunct of $Z\}$. And let $Y = \bigvee\{Y_s|\Phi(s)$ is a disjunct of $X\}$. Then if $\Phi(s)$ is a disjunct of X, $\vdash_G \Phi(s) \to Y_s$, and so $\vdash_G X \to Y$. If $\Phi(u)$ is a disjunct of Z, then $\vdash_G \Phi(u) \to -Y_s$ for every s such that $\Phi(s)$ is a disjunct of X, and so $\vdash_G \Phi(u) \to -Y$, whence $\vdash_G Z \to -Y$. ⊣

Let rank(s) = the number of necessitations $\Box E$ such that $s(\Box E) = 0$; rank(u) = the number of necessitations $\Box E$ such that $u(\Box E) = 0$.

Let *rk*(X) = max {rank(s)|$\Phi(s)$ is a disjunct of X}; *rk*(Z) = max {rank(u)|$\Phi(u)$ is a disjunct of Z}.

Lemma 3
If $\vdash_G -(X \& Z)$, then for some Y in $\mathscr{L}$, $\vdash_G X \rightarrow Y$ and $\vdash_G Z \rightarrow -Y$.

Proof. Induction on $rk(X) + rk(Z)$. Suppose that $\vdash_G - (X \& Z)$. Let $\Phi(s)$ be a disjunct of X; $\Phi(u)$, of Z. Then $\vdash_G - (\Phi(s) \& \Phi(u))$ and $\text{rank}(s) + \text{rank}(u) \leqslant rk(X) + rk(Z)$. By Lemma 2 it suffices to show that there is a sentence Y in $\mathscr{L}$ such that $\vdash_G \Phi(s) \rightarrow Y$ and $\vdash_G \Phi(u) \rightarrow -Y$.

If $s(p) = 1 - u(p)$ for some sentence letter p, we are done: Take $Y = p$ if $s(p) = 1$ and $Y = -p$ if $s(p) = 0$. So we may assume that for no sentence letter p, $s(p) = 1 - u(p)$.

Let $\Box H_0, \ldots, \Box H_{j-1}$ be all those $\Box D$ such that $s(\Box D) = 1$; let $\Box I_0, \ldots, \Box I_{k-1}$ be all those $\Box D$ such that $u(\Box D) = 1$. Then by Lemma 1, there is a sentence E in V_{su} such that $\vdash_G - (-E \& \boxdot H_0 \& \cdots \& \boxdot H_{j-1} \& \boxdot I_0 \& \cdots \& \boxdot I_{k-1})$, and therefore $\vdash_G -(-E \& \Box E \& \boxdot H_0 \& \cdots \& \boxdot H_{j-1} \& \boxdot I_0 \& \cdots \& \boxdot I_{k-1})$. Since $E \in V_{su}$, either $s(\Box E) = 0$ or $u(\Box E) = 0$. We shall assume that $s(\Box E) = 0$; the argument is similar if $u(\Box E) = 0$.

Let $J = -E \& \Box E \& \boxdot H_0 \& \cdots \& \boxdot H_{j-1}$.

Let $K = \boxdot I_0 \& \cdots \& \boxdot I_{k-1}$.

Then $\vdash_G - (J \& K)$.

By the propositional calculus, there is a sentence L such that $\vdash_G J \leftrightarrow L$, L is a disjunction of sentences $\Phi(s')$, and whenever $\Phi(s')$ is a disjunct of L, then $s'(\Box E) = s'(\Box H_0) = \cdots = s'(\Box H_{j-1}) = 1$.

Similarly, there is a sentence M such that $\vdash_G K \leftrightarrow M$, M is a disjunction of sentences $\Phi(u')$, and whenever $\Phi(u')$ is a disjunct of M, then $u'(\Box I_0) = \ldots = u'(\Box I_{k-1}) = 1$.

So $\vdash_G - (L \& M)$.

If $\Phi(s')$ is a disjunct of L, then since $s(\Box E) = 0$, $s'(\Box E) = 1$, and $s'(\Box H_0) = \cdots = s'(\Box H_{j-1}) = 1$, $\text{rank}(s') < \text{rank}(s)$, and so $rk(L) < \text{rank}(s)$. If $\Phi(u')$ is a disjunct of M, then since $u'(\Box I_0) = \cdots = u'(\Box I_{k-1}) = 1$, rank $(u') \leqslant$ rank(u), and so $rk(M) \leqslant$ rank(u). Thus $rk(L) + rk(M) <$ rank(s) + rank$(u) \leqslant rk(X) + rk(Z)$, and so by the induction hypothesis, for some sentence Q in $\mathscr{L}$, $\vdash_G L \rightarrow Q$ and $\vdash_G M \rightarrow -Q$. Thus $\vdash_G J \rightarrow Q$ and $\vdash_G K \rightarrow -Q$, and so $\vdash_G -\Diamond(J \,\&\, -Q)$ and $\vdash_G -\Diamond(Q \,\&\, K)$.

Let $Y = \Diamond Q$. Y is in $\mathscr{L}$.

$\vdash_G \Phi(s) \rightarrow Y$: For we have that

$$\vdash_G -\Box E \,\&\, \Box H_0 \,\&\, \cdots \,\&\, \Box H_{j-1} \,\&\, \Box -Q \rightarrow \Diamond(-E \,\&\, \Box E \,\&\, \boxdot H_0 \,\&\, \cdots \,\&\, \boxdot H_{j-1} \,\&\, -Q),$$

that is,

$$\vdash_G -\Box E \,\&\, \Box H_0 \,\&\, \cdots \,\&\, \Box H_{j-1} \,\&\, \Box -Q \rightarrow \Diamond(J \,\&\, -Q),$$
$$\vdash_G -\Diamond(J \,\&\, -Q), \text{ and}$$
$$\vdash_G \Phi(s) \rightarrow -\Box E \,\&\, \Box H_0 \,\&\, \cdots \,\&\, \Box H_{j-1};$$

from these it follows that $\vdash_G \Phi(s) \rightarrow -\Box -Q$, that is, $\vdash_G \Phi(s) \rightarrow Y$.

And $\vdash_G \Phi(u) \rightarrow -Y$: For we have that

$$\vdash_G \Box I_0 \,\&\, \cdots \,\&\, \Box I_{k-1} \,\&\, \Diamond Q \rightarrow \Diamond(Q \,\&\, \Box I_0 \,\&\, \cdots \,\&\, \Box I_{k-1}),$$

that is,

$$\vdash_G \Box I_0 \,\&\, \cdots \,\&\, \Box I_{k-1} \,\&\, \Diamond Q \rightarrow \Diamond(Q \,\&\, K),$$
$$\vdash_G -\Diamond(Q \,\&\, K), \text{ and}$$
$$\vdash_G \Phi(u) \rightarrow \Box I_0 \,\&\, \cdots \,\&\, \Box I_{k-1};$$

from these it follows that $\vdash_G \Phi(u) \rightarrow -\Diamond Q$, that is, $\vdash_G \Phi(u) \rightarrow -Y$. $\dashv$

The Craig Interpolation Lemma for G follows immediately from Lemma 3: Suppose that $\vdash_G A \rightarrow C$. Then for some sentences X, Z, X is a disjunction of sentences $\Phi(s)$, X is equivalent in the propositional calculus to A, Z is a disjunction of sentences $\Phi(u)$, and Z is equivalent in the propositional calculus to $-C$. So $\vdash_G -(X \,\&\, Z)$. By Lemma 3, for some Y in $\mathscr{L}$, $\vdash_G X \rightarrow Y$ and $\vdash_G Z$

$\to -Y$. Let $B = Y$. Then $\vdash_G A \to B$, $\vdash_G B \to C$, and every sentence letter contained in B is contained in both A and C.

[In what follows '$D(p)$', '$A(p)$', etc., denote sentences that may contain sentence letters other than p.]

The Beth Definability Theorem for G
Suppose that $q \neq p$, that $D(q)$ is exactly like $D(p)$ except for containing q at exactly those places at which $D(p)$ contains p, and that $\vdash_G D(p) \mathbin{\&} D(q) \to (p \leftrightarrow q)$. Then, for some sentence H containing only sentence letters contained in $D(p)$ other than p, $\vdash_G D(p) \to (p \leftrightarrow H)$.

The usual proof. By the supposition, $\vdash_G (D(p) \mathbin{\&} p) \to (D(q) \to q)$. By the Craig Interpolation Lemma for G, for some sentence H containing only sentence letters contained in both $(D(p) \mathbin{\&} p)$ and $(D(q) \to q)$, that is, sentence letters contained in $D(p)$ other than p,

(1) $\vdash_G (D(p) \mathbin{\&} p) \to H$, and
(2) $\vdash_G H \to (D(q) \to q)$. By (1),
(3) $\vdash_G D(p) \to (p \to H)$. By (2),
(4) $\vdash_G D(q) \to (H \to q)$.

Since H does not contain q, it follows from (4) by substitution that

(5) $\vdash_G D(p) \to (H \to p)$.

And from (3) and (5), it follows that $\vdash_G D(p) \to (p \leftrightarrow H)$. ⊣

We now give a lemma by means of which the more interesting half of the fixed-point theorem can be derived from the Beth Definability Theorem for G.

Lemma (Bernardi)
Suppose that $A(p)$ is modalized in p and q does not occur in $A(p)$. Then $\vdash_G \boxdot(p \leftrightarrow A(p)) \mathbin{\&} \boxdot(q \leftrightarrow A(q)) \to (p \leftrightarrow q)$.[3]

Proof. Let $E_1(p), \ldots, E_m(p)$ be p and all the subsentences of $A(p)$.

Let $B = [\boxdot(p \leftrightarrow A(p)) \,\&\, \boxdot(q \leftrightarrow A(q)) \rightarrow (E_1(p) \leftrightarrow E_1(q)) \,\&\, \cdots \,\&\, (E_m(p) \leftrightarrow E_m(q))]$.

Let $C_i = \Box B \,\&\, \boxdot(p \leftrightarrow A(p)) \,\&\, \boxdot(q \leftrightarrow A(q)) \rightarrow (E_i(p) \leftrightarrow E_i(q))$ $(1 \leqslant i \leqslant m)$.

We first show that for all i, $\vdash_G C_i$.

Suppose that $E_i(p) = \Box E_j(p)$. Then since

$$\vdash_G \Box(p \leftrightarrow A(p)) \,\&\, \Box(q \leftrightarrow A(q)) \rightarrow \Box(\boxdot(p \leftrightarrow A(p)) \,\&\, \boxdot(q \leftrightarrow A(q))),$$

$$\vdash_G \Box B \,\&\, \Box(p \leftrightarrow A(p)) \,\&\, \Box(q \leftrightarrow A(q)) \rightarrow \Box(E_j(p) \leftrightarrow E_j(q)),$$

and therefore

$$\vdash_G \Box B \,\&\, \boxdot(p \leftrightarrow A(p)) \,\&\, \boxdot(q \leftrightarrow A(q)) \rightarrow (\Box E_j(p) \leftrightarrow \Box E_j(q)),$$

that is, $\vdash_G C_i$.

If $E_i(p)$ is $\bot$ or a sentence letter other than p or q, then $E_i(p) = E_i(q)$, and so, trivially, $\vdash_G C_i$.

If $E_i(p) = (E_j(p) \rightarrow E_k(p))$ and $\vdash_G C_j$ and $\vdash_G C_k$, and $\vdash_G C_i$.

Since $A(p)$ is a truth-functional compound of necessitations and sentence letters other than p or q, if $E_i(p) = A(p)$, then $\vdash_G C_i$.

And if $E_i(p) = p$, then, as we have just seen,

$$\vdash_G \Box B \,\&\, \boxdot(p \leftrightarrow A(p)) \,\&\, \boxdot(q \leftrightarrow A(q)) \rightarrow (A(p) \leftrightarrow A(q)),$$

whence

$$\vdash_G \Box B \,\&\, \boxdot(p \leftrightarrow A(p)) \,\&\, \boxdot(q \leftrightarrow A(q)) \rightarrow (p \leftrightarrow q),$$

that is, $\vdash_G C_i$.

Every $E_i(p)$ is a truth-functional combination of sentence letters other than q and necessitations $\Box E_j(p)$. It follows that for all i, $\vdash_G C_i$, and therefore

$$\vdash_G \Box B \rightarrow [\boxdot(p \leftrightarrow A(p)) \,\&\, \boxdot(q \leftrightarrow A(q)) \rightarrow (E_1(p) \leftrightarrow E_1(q)) \,\&\, \cdots \,\&\, (E_m(p) \leftrightarrow E_m(q))],$$

that is, $\vdash_G \Box B \rightarrow B$. So $\vdash_G B$, and since $p = E_i(p)$ for some i,

$$\vdash_G \boxdot(p \leftrightarrow A(p)) \,\&\, \boxdot(q \leftrightarrow A(q)) \rightarrow (p \leftrightarrow q). \quad \dashv$$

From the lemma and the Beth Definability Theorem for G, it follows that if $A(p)$ is modalized in p, then for some sentence H containing only sentence letters contained in $A(p)$ other than p, $\vdash_G \boxdot(p \leftrightarrow A(p)) \rightarrow (p \leftrightarrow H)$, and so $\vdash_G \Box(p \leftrightarrow A(p)) \rightarrow \Box(p \leftrightarrow H)$, which is the proof-theoretically more interesting half of the fixed-point theorem.

NOTES

Introduction

1 Gödel, "Über formal unentscheidbare Sätze," pp. 173–98; the standard English translation is in van Heijenoort, ed., *From Frege to Gödel,* pp. 596–616.

2 As Quine suggests, according to Ruth Marcus (see Marcus, "Modalities and intensional languages," p. 278).

3 Lewis, "Implication and the algebra of logic," pp. 522–31.

4 Lewis and Langford, *Symbolic Logic,* p. 155.

5 Ibid., p. 23.

6 Ibid., p. 247.

7 Ibid., pp. 160–1.

8 It is curious that Lewis eventually came to regard the non-normal system *S2* as the "definitive form" of his system of Strict Implication (see p. vii of the Dover edition of Lewis, *A Survey of Symbolic Logic*). $\Box\Box(p \vee -p)$ is *not* a theorem of *S2*; a semantics for *S2* is provided in Kripke, "Semantical Analysis II," pp. 206–20, and in Hughes and Cresswell, *An Introduction to Modal Logic,* pp. 275–6.

9 In Kripke, "A completeness theorem," pp. 1–14; "Semantical analysis I," pp. 67–96; "Semantical analysis II," pp. 206–20; "Semantical considerations," pp. 83–94; and "Undecidability of monadic modal quantification theory," pp. 113–16.

10 Segerberg, *Essay,* pp. 84–8. Segerberg refers to *G* as '*K4W*'.

11 The term 'arithmetic' is sometimes used to designate the (nonaxiomatizable) theory whose theorems are the sentences true in the standard model for PA.

12 These three conditions follow from the three "derivability conditions" of Hilbert-Bernays (see Hilbert and Bernays,

Grundlagen, p. 295). Our (I), (II), and (III) were explicitly given in Löb, "Solution of a problem," pp. 115–18.

13 The construction given here is due to Quine and Smullyan (see Quine, "The ways of paradox," pp. 1–18, and Smullyan, "Languages in which self-reference is possible," pp. 55–67).

14 Rosser, "Extensions of some theorems," pp. 87–91.

15 Henkin, "A problem," p. 160.

16 Löb, "Solution of a problem," pp. 115–18.

17 The derivation of Löb's theorem from the second incompleteness theorem for single-sentence extensions of PA was told to me by Kripke.

18 As observed in Kreisel, "Mathematical logic," p. 155.

19 An effective procedure for calculating the truth-values of deictic sentences is given in Boolos, "On deciding the truth," pp. 779–81.

20 Bernardi, "The fixed-point theorem," pp. 239–51.

21 Solovay, "Provability interpretations," pp. 287–304.

22 We suppose that '$\Box\Diamond\top$' and '$\Box - \Box\bot$' express the same proposition.

23 Montague, "Syntactical treatments of modality," pp. 286–302.

Chapter 1

1 Gödel, "Über formal unentscheidbare Sätze," pp. 173–98.

2 Kripke, "Semantical analysis I," pp. 67–96.

Chapter 2

1 Kleene, *Introduction to Metamathematics*, pp. 238–41.

2 Boolos and Jeffrey, *Computability and Logic*, pp. 164–6.

3 Hilbert and Bernays, *Grundlagen*, vol. II, pp. 319–37.

4 Shoenfield, *Mathematical Logic*, pp. 211–13.

Chapter 3

1 Löb, "Solution of a problem," pp. 115–18.

2 I am grateful to Warren Goldfarb for telling me the arguments contained in the last two paragraphs.

Chapter 4

1 And thereby solve problem 35 of Friedman, "One hundred and two problems," p. 117 (cf. Boolos, "On deciding the truth," pp. 779–81). Friedman's thirty-fifth problem was also independently solved by Bernardi and Montagna and by van Benthem. An algorithm for calculating the truth-

values of deictic sentences was found by the author in the summer of 1973.

2 Jeroslow, "Redundancies," pp. 359–67.

3 Rogers, *Theory of Recursive Functions and Effective Computability*, p. 188.

4 Bernardi, "The fixed-point theorem," pp. 239–51.

Chapter 5

1 Most notably in Kripke, "Semantical analysis I," pp. 67–96.

2 Cf. Solovay, "Provability interpretations," pp. 301–2.

Chapter 6

1 In Lemmon and Scott, *Intensional Logic*, and Makinson, "On some completeness theorems," pp. 379–84.

Chapter 7

1 Segerberg, *Essay*, pp. 84–8.

Chapter 8

1 I am greatly indebted to van Maaren, "Volledigheid v. d. modale logica *L*." van Maaren refers to *G* as '*L*'.

2 Jeffrey, *Formal Logic*, p. 114.

Chapter 9

1 Cf. Bernardi, "The fixed-point theorem" (pp. 239–51) and "The uniqueness of the fixed-point," and Smoryński, "Calculating self-referential statements."

Chapter 10

1 I am grateful to Warren Goldfarb for telling me of this, Rosser's original, way of proving Rosser's theorem.

2 I am again indebted to Goldfarb for telling me the arguments contained in these two paragraphs.

Chapter 11

1 See Sambin, "An effective fixed point theorem," pp. 345–61, and Smoryński, "Calculating self-referential statements, I," for other proofs of this result.

Chapter 13

1 'Grz' is for Grzegorczyk, who (in Grzegorczyk, "Some relational systems," pp. 223–31) studied a schema deductively equivalent in *S*4 to $\Box(\Box(A \rightarrow \Box A) \rightarrow A) \rightarrow A$. Sobocinski

(in "Family $\mathcal{K}$," pp. 313–18) studied the obviously equivalent $\Box(\Box(A \to \Box A) \to A) \dashv A$.

2 Cf. Segerberg, *Essay*, pp. 96–103, from which the present proof is taken.

3 In Goldblatt, "Arithmetical necessity, provability, and intuitionistic logic" (and independently, but later, by the author; see Boolos, "Provability in arithmetic").

Chapter 14

1 For a different proof of the interpolation lemma for *G*, see Smoryński, "Beth's theorem and self-referential sentences."

2 Solovay, "Provability interpretations," pp. 293–4.

3 This is essentially Theorem 3 of Bernardi, "The uniqueness of the fixed-point."

BIBLIOGRAPHY

Bernardi, C., "The fixed-point theorem for diagonalizable algebras," *Studia Logica* 34 (1975), 239–51.

– "The uniqueness of the fixed-point in every diagonalizable algebra," preprint of the Institute of Mathematics of the University of Siena.

Boolos, G., "On deciding the provability of certain fixed point statements," *Journal of Symbolic Logic,* forthcoming.

– "On deciding the truth of certain statements involving the notion of consistency," *Journal of Symbolic Logic* 41 (1976), 779–81.

– "Provability in arithmetic and a schema of Grzegorczyk," *Fundamenta Mathematicae,* forthcoming.

– and Jeffrey, R., *Computability and Logic,* Cambridge, 1974.

Cresswell, M. J., "Frames and models in modal logic," in J. N. Crossley, ed., *Algebra and Logic,* Berlin, Heidelberg, and New York, 1975.

Feferman, S., "Arithmetization of metamathematics in a general setting," *Fundamenta Mathematicae* 49 (1960), 35–92.

Friedman, H., "One hundred and two problems in mathematical logic," *Journal of Symbolic Logic* 40 (1975), 113–29.

Gabbay, D. M., *Investigations in Modal and Tense Logics with Applications to Problems in Philosophy and Linguistics,* Dordrecht and Boston, 1976.

Gödel, K., "Eine Interpretation des intuitionistischen Aussagenkalküls," *Ergebnisse eines Mathematischen Kolloquiums* 4 (1933), 39–40. English translation in J. Hintikka, ed., *The Philosophy of Mathematics,* Oxford, 1969.

– "Über formal unentscheidbare Sätze der *Principia Mathematica* und verwandter Systeme I," *Monatshefte für Mathematik und Physik* 38 (1931), 173–98.

Goldblatt, R. "Arithmetical necessity, provability, and intuitionistic logic," *Theoria,* forthcoming.
Grzegorczyk, A., "Some relational systems and the associated topological spaces," *Fundamenta Mathematicae* 60 (1967), 223–31.
Henkin, L., "A problem concerning provability," *Journal of Symbolic Logic* 17 (1952), 160.
Hilbert, D., and Bernays, P., *Grundlagen der Mathematik,* 2nd ed., Berlin, Heidelberg, and New York, 1968.
Hughes, G. E., and Cresswell, M. J., *An Introduction to Modal Logic,* London, 1968.
Jeffrey, R. C., *Formal Logic: Its Scope and Limits,* New York, 1967.
Jeroslow, R. G., "Redundancies in the Hilbert–Bernays derivability conditions for Gödel's second incompleteness theorem," *Journal of Symbolic Logic* 38 (1973), 359–67.
Kleene, S. C., *Introduction to Metamathematics,* Princeton, 1952.
Kreisel, G., "Mathematical logic," in T. L. Saaty, ed., *Lectures on Modern Mathematics,* vol. III, New York, London, and Sydney, 1965.
Kripke, S. A., "A completeness theorem in modal logic," *Journal of Symbolic Logic* 24 (1959), 1–14.
– "Semantical analysis of modal logic I. Normal modal propositional calculi," *Zeitschrift für mathematische Logik und Grundlagen der Mathematik* 9 (1963), 67–96.
– "Semantical analysis of modal logic II. Non-normal modal propositional calculi," in J. W. Addison, L. Henkin, and A. Tarski, eds., *The Theory of Models,* Amsterdam, 1965.
– "Semantical considerations on modal logic," *Acta Philosophica Fennica* 16 (1963), 83–94.
– "The undecidability of monadic modal quantification theory," *Zeitschrift für mathematische Logik und Grundlagen der Mathematik* 8 (1962), 113–16.
Lemmon, E. J., and Scott, D. S., *Intensional Logic,* unpublished draft of initial chapters, Stanford, 1966.
Lewis, C. I., "Implication and the algebra of logic," *Mind* 21 (N.S.), (1912), 522–31.
– *A Survey of Symbolic Logic,* Dover edition, New York, 1960.
– and Langford, C. H., *Symbolic Logic,* Dover edition, New York, 1959.
Löb, M. H., "Solution of a problem of Leon Henkin," *Journal of Symbolic Logic* 20 (1955), pp. 115–18.
Macintyre, A., and Simmons, H., "Gödel's diagonalization technique and related properties of theories," *Coloquium Mathematicum* 28 (1973), 165–80.

Magari, R., "The diagonalizable algebras," preprint of the Institute of Mathematics of the University of Siena.
Makinson, D., "On some completeness theorems in modal logic," *Zeitschrift für mathematische Logik und Grundlagen der Mathematik* 12 (1966), 379–84.
Marcus, R., "Modalities and intensional languages" in I. M. Copi and J. A. Gould, eds., *Contemporary Readings in Logical Theory,* New York, 1967.
Montague, R., "Syntactical treatments of modality, with corollaries on reflexion principles and finite axiomatizability," in R. Montague, *Formal Philosophy,* New Haven and London, 1974.
Quine, W. V. O., "The ways of paradox," in W. V. O. Quine, *The Ways of Paradox and other Essays,* revised and enlarged edition, Cambridge, Mass. and London, 1976.
Rogers, H., *Theory of Recursive Functions and Effective Computability,* New York, 1967.
Rosser, J. B., "Extensions of some theorems of Gödel and Church," *Journal of Symbolic Logic* 1 (1936), 87–91.
Sambin, G., "An effective fixed point theorem in intuitionistic diagonalizable algebras," *Studia Logica* 35 (1976), 345–61.
Schütte, K., *Vollständige Systeme modaler und intuitionistischer Logik,* Berlin, Heidelberg, and New York, 1968.
Segerberg, K., *An Essay in Classical Modal Logic,* Uppsala, 1971.
Shoenfield, J. R., *Mathematical Logic,* Reading, Mass., 1967.
Smoryński, C., "Beth's theorem and self-referential sentences," forthcoming.
– "Calculating self-referential statements, I: explicit calculation," to appear in the Proceedings of the 1977 ASL colloquium at Wrocław.
Smullyan, R. M., *First-order Logic,* Berlin, Heidelberg, and New York, 1968.
– "Languages in which self-reference is possible," *Journal of Symbolic Logic* 22 (1957), 55–67.
Sobocinski, B., "Family $\mathcal{K}$ of the non-Lewis systems," *Notre Dame Journal of Formal Logic* 5 (1964), pp. 313–18.
Solovay, R., "Provability interpretations of modal logic," *Israel Journal of Mathematics* 25 (1976), 287–304.
van Heijenoort, J., ed., *From Frege to Gödel: A Source Book in Mathematical Logic, 1879–1931,* Cambridge, Mass., 1967.
van Maaren, H., "Volledigheid v.d. modale logica *L*," Thesis, Mathematical Institute of the University of Utrecht, 1974.

INDEX

accessibility relation, 74
ancestral, 77
antisymmetric relation, 160
appropriate to *G*, 81–3
appropriate to *K* (*T*, *K*4, *S*4, *B*, *S*5), 80–1
arithmetic, 5, 35
arithmetization, 52
available occurrence, 113

B, 23
Bernardi, C., 13, 132, 173
Bernardi–Smoryński theorem, 57, 66
beta-function, 38
Beth, E., 108
Beth definability theorem for *G*, 173

canonical model, 87
Carnap, R., 89
characterize, 83
closed, 113
composition, 37
Craig interpolation lemma for *G*, 168–73

de Jongh, D., 18, 30, 141
decidable formula, 54
degree (of nesting), 112
deictic sentence, 61
derivability conditions (Hilbert-Bernays), 6, 166
describe, 83–4
distribution axiom, 22–3

equivalent, 35
euclidean relation, 80
evaluator, 74
extend, 23

finite strict partial ordering, 99
finite weak partial ordering, 160
fixed point, 7, 50
fixed-point theorem (de Jongh–Sambin), 87, 141–6
frame, 74

G, 2, 19, 23
*G**, 14, 57
generalized diagonal lemma, 49
Gentzen, G., 108
Gödel, K., 1, 13, 14, 19, 40, 140
Gödel numbering, 6
Gödel sentence, 13, 65–6, 123
Gödelian fixed point, 13, 66, 123
Gödel's first incompleteness theorem, 53–4, 133
Gödel's second incompleteness theorem, 53, 133
Goldblatt, R., 161

Henkin, L., 10
Henkin sentence, 13, 65–6, 123

identity function, 37
identity term, 39
imply, 35
incomplete theory, 7, 53
inconsistent theory, 9

Jeffrey, R., 115
Jeroslow, R., 65
Jeroslow sentence, 65–6, 123

K, 23
*K*4, 23, 47
Kreisel, G., 138
Kripke, S., 4, 11, 30, 72, 73, 74, 89, 108

L-consistent, 90
letterless sentence, 21
Lewis, C. I., 2, 4, 15
lie, 62
Lindenbaum, A., 89
Löb, M. H., 10, 11, 13, 14
Löb's theorem, 10–11, 54

Makinson, D., 85, 89
maximal, maximal (*L*-) consistent, 91
modal sentence, 20–1
model, 74
Montague, R., 16–17

necessitation (of a sentence), 21
necessitation, rule of, 22
normal form, 62
normal system, 5
numeral, 6, 37

1-consistent theory, 54
open, 113

Peano Arithmetic (PA), 5, 35
possible world, 74
predicate, 35
prim rec list, 40
primitive recursion, 37
primitive recursive formula, 44
primitive recursive formula in the strong sense, 43
primitive recursive function, 37
primitive recursive relation, 37
primitive recursive term, 39

Quine, W. V. O., 15–16

realization, 11, 46
reflection principle, 55
represent, 37
Rogers, H., 66
Rogers sentence, 66, 123
Rosser, J. B., 8, 134
Rosser sentence, 138
Rosser's theorem, 133–7
rule (□, – □), 113

*S*4, 23
*S*4*Grz*, 160
*S*5, 23
Sambin, G., 18, 30, 141
Scott, D., 87, 89
Segerberg, K., 100, 161
Σ_1-formula, 44
Σ_1-formula in the strong sense, 44
Σ_1-sentence, 44
Smoryński, C., 13, 132
Smullyan, R., 108
Solovay, R., 14, 85, 97, 168
Solovay's Completeness Theorem for *G*, 14, 51, 151
Solovay's Completeness Theorem for *G**, 15, 58, 152
standard model for PA, 10, 36
strict partial well-ordering, 83
subsentence, 22
substitution, 22
successor function, 37
successor term, 39

T, 23
term composition, 39
term recursion, 39
translation, 11, 46
truth-translation, 159

undecidable sentence, 7, 53

valid sentence, 76
validity, in a frame, 76
  in a model, 76

well-capped relation, 83
well-founded relation, 5, 82
world, 74

zero function, 37
zero term, 39